THE
HUNT
FOR
BIGFOOT

Revised & Updated

CRYPTO

editions

ISBN-13: 978-0-88839-113-1 [trade paperback]
ISBN-13: 978-0-88839-453-8 [epub]
Copyright © 2021 Peter Byrne
Crypto Editions
an imprint of Hancock House Publishers

CRYPTO
editions

Library and Archives Canada Cataloguing in Publication

Title: The hunt for Bigfoot / Peter Byrne.
Names: Byrne, Peter, 1925- author.
Description: Revised & updated.
Identifiers: Canadiana (print) 20210107006 | Canadiana (ebook) 20210107014 | ISBN 9780888391131
(softcover) | ISBN 9780888394538 (EPUB)
Subjects: LCSH: Sasquatch.
Classification: LCC QL89.2.S2 B97 2021 | DDC 001.944—dc23

Cover Painting: by Seth Jacobs, of Les Cerqueux, France

Cover Photography: Jack Sullivan, Pacific City, OR

We acknowledge the financial support of the Government of Canada through the Canada Book Fund and the Canada Council for the Arts, and of the Province of British Columbia through the British Columbia Arts Council and the Book Publishing Tax Credit.

Hancock House gratefully acknowledges the Halkomelem Speaking Peoples whose unceded traditional territories our offices reside upon.

Published simultaneously in Canada and the United States by
HANCOCK HOUSE PUBLISHERS LTD.
19313 Zero Avenue, Surrey, B.C. Canada V3Z 9R9
#104-4550 Birch Bay-Lynden Rd, Blaine, WA, U.S.A. 98230-9436
(800) 938-1114 Fax (800) 983-2262
www.hancockhouse.com sales@hancockhouse.com

THE
HUNT
FOR
BIGFOOT

Revised & Updated

PETER BYRNE, F.R.G.S.

TABLE OF CONTENTS

Dedication

This book is dedicated to all of my stalwart and redoubtable Bigfoot hunting companions of the years. Spread across nearly six decades, I hope that I can be forgiven when I say that they are now just too numerous for me to single out and name individually.

But those of you who read this will know who I mean and to whom I am grateful, for steadfast companionship in the field under all kinds of rugged conditions, for unwavering friendship, for staunch and unshakable support of my fumbling efforts to find the elusive quarry of our great mutual quest, plus my ever-striving scribbling, through three books, to make and keep an accurate record of our findings, our beliefs and our meaningful time together.

Among them now are not a few still living but, sadly, too numerous with the passage of the years, many who have gone on the last safari.

#####################

Introduction

This book, *The Hunt for Bigfoot: revised & Updated* can probably be regarded as two books in one, in that it contains most of the text of my original book, *The Search for Bigfoot*, published in 1975, plus considerable additional new writing under the current title. The new work is preceded by this Introduction and concludes with an Epilogue which brings my experience with the phenomenon, and my research, current to the present day. As to the original work, some small changes have been made. These are mainly the deletion of outdated materials; otherwise the 1975 edition-now out of print-is included as it was originally published.

The 1975 version was written by me in twenty-eight days, tapped out at two-fingered speed on an ancient Olivetti portable typewriter while in temporary residence in a twenty-five-foot trailer in a two-bit trailer court in the bleak little Columbia riverside town of The Dalles, in Oregon. The reason I was using an old, borrowed, somewhat drafty trailer at that time was that, having just moved out of lodgings in The Dalles and being in waiting for a new trailer of my own to arrive from Portland, I needed a place to stay. The trailer was thus a temporary loan. And the reason I wrote the book was a suggestion that I do so by an old friend, the quite famous, British Loch Ness monster hunter, Tim Dinsdale, who was visiting me at that time from the UK and who was surprised, he said, that until then I had not published anything about the phenomenon other than magazine articles.

Tim had just published his own book, *Monster Hunt* (1972), about the Loch Ness monsters and it was he who found a hardback and paperback publisher for me, one Alphonse J. Hackl, the owner of Acropolis Books, in Washington, DC. He also introduced me to a Boston-based patent attorney, Robert Rines, who wrote the Introduction to the work and who, to the time of his death in November, 2009, personally sponsored much of my Bigfoot research work. Bob Rines was a lifelong friend.

Hackl told me that he sold copies of what was, if nothing else at that time, an unusual book, to every library in the US and that in all he was able to dispose of about 2000 copies. Then the book, as do most publications, lapsed into literary obscurity and disappeared from the shelves of book shops. Recently, however, with the renewed awareness of the possibility of the reality of the Bigfoot phenomenon, there has been a resurgence of interest in positive writings about the mystery, as a result of which I have received a large number of email inquiries about the 1975 book, along with many old and battered copies of it being sent to me for signing and inscription.

This renewed interest in a work that was the first book on the subject of the Bigfoot mystery by a major American publisher, is what has prompted me to republish the 1975 edition. In addition to this, I feel it is important, now, after five decades of investigation and research into the phenomenon, that readers be given something of an in-depth look at its whole broad spectrum, one that includes its early history—as far back as the eighteenth century—and that goes on from there to when it first came to the attention of the public in 1959, one that is rounded off, now, 2015, with an account of its present status in both public and scientific domains.

The Bigfoot mystery came to the attention of the world in 1959, mainly through the interest of a wealthy and adventurous Texan, a man named Tom Slick, who, in that year, on the basis of personal interest and healthy curiosity, launched the first investigations in northern California. He started with a rough-knit research group that operated in California and British Columbia. The group had no headquarters, each member operating from his own home, and it consisted of a Canadian free-lance journalist, a Swiss Canadian lead scavenger-mostly used shotgun birdshot-a truck driver, an enigmatic taxidermist, and two odd gentlemen of somewhat dubious character who described themselves as "hunter trackers."

The group was a diverse collection, to say the least. But, as Tom explained to me later, when he first became aware of the phenomenon and traveled to the Pacific Northwest to investigate it, these were the only people he could find who had any interest in it, or at least who said they did. US Forest Service and US Fish and Wildlife personnel with whom he talked dismissed the subject with a wave of the hand, so to speak, as did State Police and Sheriffs' departments. So, until such time as he could bring in a professional research team from San Antonio, with all of its logistics and expenses, which he planned on doing after he got enough evidence to support and satisfy his own curiosity and warrant a major search, this seemed the best way to go. And so he put together a meeting with the group, told them what he wanted and to that end drew up a simple contract, one, he told me later, that he thought best suited to the intellectual levels of the individuals in question.

The contract stated that their work would consist of general research into the phenomenon in northern California and British Columbia and that it would include field work and inquiry into incidents such as sighting claims and footprint finds. Each member of the group was required to make a report to Tom once a month, by regular mail. Even if there were no findings, this routine check-in was required, if only to assure Tom that basic research work was ongoing. In return for this, each was to be paid an honorarium of $300 a month plus general expenses to cover such things as gas, basic camping supplies, and food while in the field. Tom would run the operation from his offices in San Antonio, Texas, and collect calls, from any of the team, would be accepted at all times.

When he put together the first phase of his Bigfoot searches, I was in the Himalaya, working through the third year of the three-year Yeti search expeditions, which he and his partners Kirk Johnson, of Texas, and Jimmy Stewart, of Beverly Hills, were sponsoring. These came to an end in December, 1959, about which time, by coincidence, after about six months of desultory research, Tom's Pacific Northwest Bigfoot project was beginning to show signs of collapse.

The Himalayan Yeti expeditions were planned and budgeted for three years, 1957, '58 and '59, and that is how long they lasted. Tom's Pacific Northwest Bigfoot project did not have a schedule; its duration was to be determined by how its team performed and what results they produced and contributed, which were to be passed on to Tom via the monthly letter or, in the case of a serious find of some kind, via a collect telephone call. When, after an initial, promising enough start, the monthly reports became fewer and fewer, and eventually ceased, Tom became concerned. He became even more so when he learned through the grapevine—which may have been created by the American members of his team—that there was childish but nevertheless quite vicious infighting going on between the two Canadian members of the group, the journalist and the lead scavenger. Adding to his concern was the fact that his two hunter-tracker team members seemed to have completely disappeared, along with a $4500.00 grant he had given them to enable them to set up a Bigfoot research base camp in the mountains of the Six Rivers National Forest and spend six months in it. The combination of these problems left him with a difficult decision.

Tom did not want to close down his Bigfoot investigation. He was intrigued by the possibility of the reality of a large, unclassified hominid living in the vast forests of the Pacific Northwest and quite convinced that dedicated research would in time bring about a finding and, with it, the exciting prospect of communication with the creatures. This great interest in the phenomenon, which he maintained until the day he died, was at this time all the more encouraged by the historical research of a fine Native American journalist, a Ms. Betty Allen, of Willow Creek, California, whose work on behalf of the project was beginning to unearth not only an Indigenous background to the phenomenon but also legends and stories—many of them recorded in old books and newspapers—dating back to the 1700s.

So Tom dismissed the idea of shutting down the project; that was not on the books for the moment. But if his goals were to be reached, he had to get a productive and conscientious team together and get them into the field and engaged in serious research. The answer was fairly obvious. Get rid of the bulk of the team he had now—with its stipend-squandering, heel-dragging, uncommunicative, infighting members, in particular the Canadian troublemakers—and set up a new team under new leadership. And in December, 1959, he made a decision to do just this.

In the little Kingdom of Nepal, in the 1950s, communication was very limited. Within the country there was no phone, postal or telegraph service and if one wanted to send a message, anywhere, a porter had to be engaged to carry it, on foot, for delivery, something that-depending on the distance involved- could take weeks. Two international cable stations existed, however, at the only two embassies operating in the country at that time. These were the British and Indian embassies in Kathmandu, and of the two, the Indian one allowed public access for receiving and sending cables; the British, with stiff upper lips, did not. So when we were in Kathmandu—which was once a year for Christmas— we used the Indian embassy for our cables, both sending and receiving and in December, 1959, in the ancient city for the festive season and the termination of our Yeti expeditions, one of the many cables we received was from Tom Slick. What he wrote, I am embarrassed to say now, at first made us laugh; we truly thought he was joking.

It was mid-morning of December 31st, the last day of the year, and my brother Bryan and I had just arrived in the city—after twelve months of tent life and cave life in the high Himalaya—via a six-day, two-hundred-mile hike from the Sola Khumbu District. We were very dirty, bearded, with ragged clothing and worn-out boots and gear. But we were sitting back in big canvas chairs in the Yak and Yeti Bar of Katmandu's only foreign visitor hotel, the Royal, and were doing what any explorer would do at the tail end of three years of hard yakka—as the Aussies call it—in rugged, high mountain country, sitting back and eating delicious curried bar snacks and drinking cold beer. And also reading our mail, a huge pile of it, one that also contained cables of congratulation for completion of the three-year set of expeditions from Tom, Kirk Johnson—his partner—and many others.

The cable from Tom, which is recorded in journals to this day, was brief enough. It read:

Congrats on excellent work and completion of investigation. Have conferred with Kirk Johnson and others and we feel that you should now take a break. If you agree, we would like you to come to the U.S. and help us investigate the American Yeti, a group of which our research suggests live in the forests of the Pacific Northwest. Take a break, enjoy the Royal for as long as you like—at our expense—and get back to us when you have a decision. Tom.

I read it twice and then handed it to Bryan. He read it twice, too, and then looked at me with eyebrows raised. "A what? In the north somewhere forests?" He said. "A Yeti? Is Tom OK? Or is he joking? Where are these forests? How big are they? Certainly not as big as where we have just spent three years. Must be west of New York or somewhere like that. Maybe the west coast? Would that be California? Or Oregon. But a Yeti? And how did it get there? From Nepal? In which case what airline did it use?"

As I mentioned, for a while we thought Tom was joking. But I knew the man quite well by then, having spent time with him in both India and the Himalaya, and while I knew he had a sense of humor, I also knew that he was a highly intelligent and serious person, one not given to vain statements or frivolous pursuits. So next day I cabled him back via the Indian embassy and acknowledged his wire and asked for more details. Did the American Yeti have a name? Was there any background history? Were there current sightings and footprint finds? How big were the things and what were was believed to be the general area of habitat?

In reply, a few days later, we got another cable from Tom, this one with a lot more detail. The things—the creatures that he had named the American Yeti—were called Sasquatches by the Pacific Northwest Indigenous. They were big—much bigger than the Himalayan Yeti—their habitat was the forests of northern California, Oregon, and Washington, and there were, apparently, many credible sightings. The cable went on with considerable detail, all of which we read with fascination. And what we realized, from what Tom had written, was that he was deadly serious about the phenomenon and about the possibility of its reality.

When I had first teamed up with Tom Slick, in 1956, I was a professional big game hunter, operating my own outfit in the jungles of far southwest Nepal and running trophy safaris for tiger, leopard, buffalo, bison, croc, wild boar, and deer. At that time I had been engaged in this business—after a five-year, self-imposed hunting apprenticeship in the forests of north Bengal, while a Tea Planter there with a British company—for three years. I gave it up to join Tom in the Himalayan searches and now I had a decision to make, which was whether to resume my hunting career or put it aside for another stint at Yeti hunting, this time for one that supposedly lived in North America.

I thought about it for a couple of days—it was a big decision for me and I was not at all sure about what hunting the American Yeti had to offer—and then cabled Tom and told him that I would come to the US and help him set up a project but that I needed a break for two safaris I had promised two sets of clients and also to visit my parents, who had recently moved to Australia and whom I had not seen in three years.

Tom's reply was both succinct and sensitive to my request. It basically said, take your time, come when you are ready, all travel costs on us here, advise air ticketing needs when the time comes, enjoy the break, keep in touch. But it added something very interesting. This was to the effect that if I had time before the safaris were due, would I be interested in joining him and his lady friend, Cathy McLean, in Delhi, for some research into one of his personal interests, what he called "mind science," which same would involve a journey overland by train and by car from Delhi to Bombay and might take about a month. As the research would include visiting Indian villages where no one spoke English, my knowledge of Hindustani would be very useful to the work. He added that he was on his way to Delhi within the week, with the trip planned to start as soon as possible.

I tossed around Tom's Indian proposal in my mind. I liked the idea of the Indian road trip and asked Bryan if he would like to join me. He said no; he had made plans to go back to Ireland and he left a few days later. I cabled Tom and said yes, that I would come with him and Cathy on their Indian trip and a few days later, with our Sherpas all paid off and train ticketed for the trip back to their homes in Darjeeling, loaded with all of our remaining camp gear which we simply gifted to them, I jumped on an Indian Airlines flight and flew to Delhi to join Tom and Cathy.

Tom's mind science interest was something that he and I had talked about several times. The foundation of it was his belief that the human brain was capable of far more than was normally realized, with a potential far beyond the present knowledge of science. India, he had concluded from reading and study, was a land where, it was claimed, *fakirs*—holy men—and others were capable of mind usage that enabled them to perform mind over matter feats which we in the West would look upon as miraculous. The trip he planned would be to investigate some of these claims, which included the Indian Rope Trick—where a man could climb up a rope that had no upper attachment and then disappear; the ability to walk on water; the capability of a single man, unassisted, to lift enormous weights; and the capacity—in this case that of a Bombay nun—to produce and exhibit physical objects, such as charms and holy medals, literally out of the air.

Tom and Cathy and I left Delhi by train on January 5th, 1960, and traveled, by train and by car, south, for a month. Fascinating stops along the way, by invitation, were visits at the extraordinary palaces of the Maharaja of Baroda and his family, the Maharaja of Burdwan, and the wealthy, Parsee, Modi family of Bombay, all of whom Tom had contacted in advance via his newly established Mind Science Foundation in San Antonio, Texas. We also visited many ashrams, small villages and shrines and in them interviewed many a fakir, holy man and woman. Alas, not a single one of them, even on the promise of generous monetary award, was able to perform a single one of the extraordinary miracles attributed to them by the public and by their peers.

The holy man who was supposed to be able to lift a 200-pound stone with one hand did so. But only with the help of half a dozen of his disciples and their twelve additional hands. The fakir who was reputed to be able to walk on water gave it a try for us, but sank. The seer who was supposed to be able to perform the Indian Rope Trick told us that his father had done it many times, but died before he was able to teach him how to do it. And the holy nun who could produce objects out of the air seemed to be able to do so—with the cleverest sleight-of-hand artistry that we had ever seen—until a high-speed camera operated by a professional cameraman, hired by Tom for the job, showed us, in slow motion, how and where she kept the objects…in her hair, in her mouth and in her ears.

We arrived in Bombay on February 1st and stayed for a few days with the delightful, aforementioned Parsee couple, Piloo and Vina Modi, in their splendid home, high above the Bombay seafront on Malabar Hill. Then we broke up, Tom and Cathy returning to the US and I flying to Sydney, Australia, to visit my parents. Tom's last words to me were that he wanted me to come to the US, to help him find a Sasquatch, and that I should take my time but definitely plan on coming. I promised I would and after eight delightful weeks in Australia—one month of which was spent with my father on a hunting and fishing trip in the magnificent Australian outback in a VW camper van—I flew to New York.

There, bunked up as Tom's guest at the sumptuous Plaza Hotel, I spent a few days meeting with some of Tom's associates, including his New York secretary, Helen Schuberg. Then I flew to San Antonio, Texas.

In San Antonio, settled in as a guest at Tom's beautiful city home, I sat down with him to formulate a plan for his Pacific Northwest project. One of the first things that Tom had me do was to spend time studying maps of the Pacific Northwest and what I saw in them—a rugged area three times the size of the Himalayan ranges where I had just spent thirty-six months—impressed me greatly. There were vast stretches of forest, dozens of mountains, many of them more than 10,000 feet, hundreds of lakes and rivers and a rugged coastline with deep inlets, especially in British Colombia. After Tom told me that the whole area had ample food for a creature which, like the thousands of bears that were known inhabitants of its forests, might well be omnivorous, I realized that the terrain I was looking at was completely in line with the equation essential to the survival of a large, shy primate, a formula the critical factors of which were food, water, cover, space, the basic requirements of just about every creature on the face of the earth, from a mouse to an elephant. In other words, the creatures, if they existed, had everything they needed to both survive and to stay elusive in what was for them a perfect habitat.

I stayed in San Antonio for a week, enjoying the gracious hospitality of Tom and his family and his lady of that time, Jeri Walsh, and discussing the problems of his present project and what should be done with it, the gist of which was that I was to proceed to the Pacific Northwest, set up a base, acquire a new team and get a research project going. I was to have Hire & Fire authority with the old team and its members and I was to act on that as a primary concern, if only to relieve Tom of the thousands of dollars he was still putting out in honorariums and expenses.

From San Antonio I flew to Monterey to pick up a Jeep that was being donated to the project by a friend of Tom's. From there I drove north—remembering to drive on the right (!)—to Willow Creek, California, which Tom had chosen as the best location for a base for the new project, his choice being based on quite credible evidence, old and new, that this general area had produced and was supposedly still producing in the form of sightings and footprint finds.

Tom had one contact in the Willow Creek area at this time and this was the aforementioned Betty Allen and when, after spending a couple of nights at the Deep Sleep Motel in the nearby Hoopa Indian Reservation, I told her that we needed a house as a base for the new project, she put me in touch with a local couple, Bill and Elsie Oden, who had a two-bedroom house with space for an office, for rent at a place called Salyer—a scattering of houses—on the Redding Road about a mile south of Willow Creek. I went to see the Odens, inspected the house and, finding it suitable, as well as close to a small grocery store cum post office, moved in immediately.

My next job was to locate and execute my Hire & Fire mandate from Tom on his erstwhile group of researchers. I was to meet with them or contact them by phone, find out what their intentions were, why they were not living up to the terms of their contracts with Tom and, if their explanations were not satisfactory, to let them go. The order was to apply to all of them with the exception of the taxidermist who, Tom told me, had a special project which put him in a different category from the others, one that excused him from regular reporting and routine field work. It was a separate project that the man had personally designed and privately operated; it was also so delicate that it was being kept secret from the other members of the team. In fact it was so confidential and, if it was genuine, so promising, that Tom regretted it but said that he could not even share it with me at the moment. Later, maybe. But not now.

Moving ahead with my mandate for the five other group members was, for three of them, easier than expected. Tom had given me phone numbers for these three and had advised them all in advance that I would be calling them to discuss reorganization of the team. The remaining two, the so-called hunter trackers, I would have to track down myself; he had not had a word from them in months, or any accounting of their $4500.00 "mountain base camp" building grant.

The first team member I talked to was the Swiss Canadian lead pellet salvage man and he made it quite clear, in a guttural Swiss accent and in rapid-fire broken English laced with virulent invective—that would, as the song goes, make a sailor blush—that he was not going to work with "some frigging newcomer who probably wouldn't know a frigging Bigfoot from a frigging Black bear." It was with considerable difficulty that I managed to squeeze in a few succinct words, to the effect that he was fired.

The next one that I talked to by phone, or tried to talk to, was the Canadian journalist. His wife answered the phone and when I introduced myself she made my job relatively easy by saying, "My husband is not going to work with you and does not want to talk with you, goodbye." I started to say, "In that case we will have to let you go," but the lady hung up and that was that.

Number three was the truck driver, also by phone. He had enjoyed working with Tom, he said, and regretted that he had been unable to keep up with the research. But, he explained, he had recently moved to San Francisco and was tied up with a new job there, which made research for the project, and for Tom, impossible. He was tendering his resignation from the first of the month and expected his honorarium to cease from that date. I told him that this was acceptable and that he was free to go. He seemed like a pleasant young man on the phone and this proved to be true when we met, years later, and spent some congenial days together in the Six Rivers National Forest, sharing campfire stories about Bigfoot.

Three down and two to go. But whereas the first three had been quick and easy, the two itinerant hunter trackers—gentlemen with what appeared to be no fixed place of abode—proved to be a task of a different dimension. In fact, it took me all of three months, to the end of August, before I got word of where they were or, as it turned out, where they had recently been.

The first lead I got on the two men was from a large, smiling, and very helpful Hoopa Indian lady in a grocery and general merchandise store at a place called Witchepec, close to the Klamath River, on the edge of the Hoopa Reservation. "Yes," she said, "I remember those two and the reason I do is that they bought a huge amount of groceries here, really a lot of stuff, and said they were going to make a Bigfoot hunting camp up in the mountains. In fact they told me where they were going to locate it, a place called Doctor Rock, up in the Six Rivers Forest, north of here."

I acted promptly on the information and two days later, after a hot, ten-mile hike from the nearest motor road, accompanied by my brother-in-law, Bill Green, who was visiting from England, I found the men's camp. It was located in a small grove of trees at the edge of a big meadow, close to the place the store lady had named, Doctor Rock. It was deserted and the signs indicated that it had been so for many weeks. It also presented, to my astonished eyes, one of the most extraordinary sights I had ever seen in many years of camping.

The Pacific Northwest has a large bear population, with numbers in the tens of thousands. The animal is the common Black bear, a fairly harmless and timid animal quite different from its northern cousins, the giant Brown and the grizzly. Wounded, or cornered, the Black can be provoked and may be dangerous; females with cubs should always be regarded so. But otherwise the animal is generally not considered a threat and if in contact with man, as happens in the northern winter months when food is scare in the forests and often under deep snow, its worst habit is usually coming into town at night and getting into garbage cans, to which end it has a nose that can pick up the odors of trash from a great distance. It will apply this same ability to camping supplies and seasoned campers in the Pacific Northwest will know of this and will take appropriate precautions, which will include putting food away at night, or, if they leave camp for a short time, making sure that it is either in metal bear-proof containers or strung up in trees and high enough off the ground that it is beyond their reach. This is what the Doctor Rock so-called hunter trappers had done with their food and camping supplies for their Bigfoot base camp and it appeared that they had brought along cloth bags for this purpose. The mistake that they made was they simply did not hoist them high enough.

The Black bear is a comparatively small animal, males running a maximum of about four feet, nose to tail. And this is what our departed campers had taken into consideration when they hung their gear up in the trees around their campsite. What they did not think about is that bears have hind legs and that standing up on these will increase their body length another two feet, as well as which they also have front legs which will allow for as much as another two feet and, thus, the ability to make successful swipes at something as much as six or even seven feet above the ground. The result, for the gear and food that our two "experts" had hung four to five feet above the ground, was total disaster, or, if you like, a once-in-a-lifetime teddy bear's picnic.

The meadow, on the side of which their camp had been located, was about an acre in extent and thanks to the happy bruins' picnic it was strewn, literally end to end, with the remains of hundreds of items of half-eaten food and camp gear, every single piece of which had been partially eaten, or torn apart, or ripped to shreds, or, in the case of hundreds of cans of food, simply crunched up and sucked dry.

There were shredded down sleeping bags and down-filled pillows, chewed-up canvas camp chairs, ripped blankets, torn cotton shirts, split Levi trousers, holed gumboots, pieces of woolen socks, slashed plastic rain gear, leather boots, plastic sandals, ruined tools—hatchets, axes, and shovels—with their wooden handles chewed and splintered, the shredded remains of a large green canvas tent and the totally trashed remnants of hundreds of different kinds of food, in cans and paper and plastic and cardboard packages, edibles that included beans, soup, meat, peas, tomatoes, potatoes, onions, chili, corn, corn beef hash, waffles, carrots, biscuits, bread, flour, sugar, salt, oatmeal, and cereal. The only food that we found that had not been eaten was coffee beans and black pepper; the bears had bitten packages and cans of these into two or three pieces but, finding the contents not to their taste, had abandoned them.

We stayed at the campsite for about an hour and then hiked out and drove back to Salyer. The next day, Bill left to return to England.

Thereafter, however, whenever we shared a telephone call, invariably one of us would ask the other, with a chuckle, "Got any canned food to spare, over there?"

That was the amusing side of it. The more serious side was, where were our two hunter trappers? What had happened to them? Were they even still alive? A question which I carried with me for another month until, as so often happens in quests like this, a chance conversation with a stranger—in this case a police officer—gave me the answer.

In Willow Creek there was a tavern—I forget its name now—on the south side of the main street opposite the gas station and I went there from time to time of an evening to have a drink. One evening, I was sitting at the bar when a deputy sheriff walked in and joined me. He nodded to me as he sat down and a few minutes later introduced himself and asked me if I was the Britisher who was looking for Bigfoot. I told him yes, and as we talked about the phenomenon it was interesting to see that, as a Northwest native, he had a very positive view about the reality of the Bigfoot phenomenon. But it was when he asked me who was working with me in the project that I realized he might just know something about the two missing members of the 1959 team. So I asked him, giving him their names and telling him about their association with us, the ruined camp at Doctor Rock, and our missing funds. When I did he looked at me with a smile and said, "Oh, those two? Do I know anything about them? You bet. We have 'em in jail in Eureka. Awaiting trial. Took us six months to track the buggers down…following a paper trail of bounced checks all across northern California. That's where they are now. You wanna come visit them?"

Tom's Hire & Fire mandate now completed—with the exception of the taxidermist and his secretive project—it was time to move ahead with the general work of the project. I started by setting up an office in our rented house at Salyer. I bought and installed a desk, two office chairs, a telephone, a steel filing cabinet, a large electric Olivetti typewriter and all kinds of paper supplies. I rented a mailbox at the nearby post office and opened a credit account with their marketing section. I bought a big, used Ford station wagon—big enough for two people to sleep in with the back seat folded forward—and this, along with our donated Jeep and two Tote Goats—small, single-seat, 25 MPH motorcycles, purchased in Eureka—composed our Bigfoot mobile fleet.

My younger brother, Bryan, his Himalayan days now over and looking for a job, flew from Ireland to join me, bringing with him, from New York, a delightful and beautiful, tall, black-haired German girl, Antje, who stayed with us for a happy two years and then left us to marry a San Francisco boat builder. Last heard of, she was living in Kashmir, India. And I personally was joined by a delightful lady friend of many years, one Shirley Laurence, from Dublin, Ireland, whom I had last seen in Nepal, where, for a while, she had accompanied Bryan and me, Yeti hunting, in the mountains. Like Antje, she also eventually left us to marry a San Francisco-based advertising agent, one Ron Ward. At last word (2015), she is still married to the same man and living happily in the San Francisco Bay area under the name of Reva Ward.

On arrival at Salyer, Shirley and Antje quickly became friends and together took over the project office, thus relieving Bryan and I of the unwelcome chores of paperwork, filing and phoning and mail and monthly reporting and bill paying.

Soon after my brother's arrival I hired three men. The first of these was a local man, a semi-retired road construction worker named Gerry Crew, who had been recommended to me by Betty Allen as being trustworthy and reliable and, most important, very knowledgeable on the topography of the local area. Gerry worked with me for two years and proved to be capable, dependable, and not averse to hiking many miles with us in the mountainous country and putting in many a twelve-hour day under very often tough conditions of adverse weather and rugged terrain.

Soon after Gerry came on the team, his nephew, Jim, also joined us and between stints at college, studying for a degree in biology, spent much of 1960 and '61 with us, proving himself to be, like his uncle, a young man of energy, zeal, and good character.

The third man to join my team was one Steve Matthes, who lived in Pasa Robles, California, and was at that time coming to the end of a twenty-five-year career with US Fish & Wildlife as a professional mountain lion hunter. If I recall correctly, at the invitation of Tom Slick to join the team, he took a year's sabbatical from F & W; he also moved, about this time, from Pasa Robles to Carlotta, a town on the California coast not far south of Eureka, a move that enabled him to spend a lot more time with us at Salyer.

Matthes' job with F & W meant that he'd spent his life cropping lions. To this end, he'd developed and bred a pack of hunting hounds and he used these to chase the lions, tree them, and shoot them from the ground. He worked from a small, one-man trailer which he pulled behind an old Ford pickup and, a real outdoors man who truly loved the NW forests and his work, he often spent a month at a time camping alone with his dogs.

Matthes was a big man, six feet plus, hard-muscled and a veritable powerhouse of energy, as I soon found out when I started hiking with him up and down northern California's bone-cracking hills and through its deep gorges and river ravines, with their near impassable barriers of house-sized boulders and mountainous tangles of fallen Douglas fir trees. He was also a first-class naturalist, an expert tracker and a genial campfire companion, the latter something that soon developed and eventually cemented between us a friendship that lasted for many years.

Early in 1960, we got under way with our research. I recruited a team of volunteers, all local and well-established people from Willow Creek and the Hoopa Reservation, advertised our presence—and need for information—in a local newspaper and settled in to what was really something quite unique…a group of people engaged in a project to find a giant unknown and unclassified primate in a vast and rugged terrain that stretched from northern California to the southern border of Alaska, a project that would include historical research, investigation of finds and stories of all kinds, and many, many hours spent in the basic field work of searching for physical sign of our quarry, mainly, of course, for footprints.

We did find a few; some of them were quite fresh and looked to me and to Matthes to be genuine. They were human in shape and definition— e.g. five toes—but much too large and with too deep an impression to have been made by an average person.

In addition, we interviewed many people who claimed to be eyewitnesses, not a few of them seeming to be quite credible. We also, working with Betty Allen, dug up a mound of historical data, all of it, without exception, describing exactly the same thing…a large, hair-covered primate, bipedal (upright-walking), shy, inoffensive, elusive, secretive, and given to living alone. And this is what did, and how the project progressed, through 1960 and '61 and into 1962.

Much of the research of the project obviously bears documenting. But to do so here would be to repeat what the following pages—the book itself—already contain, which is pretty much a complete account of the work. However, one story which is not contained in the book bears telling, if only because it accounts for the last of the 1959 Bigfoot brigade—the enigmatic taxidermist—and at the same time hopefully satisfies the reader's curiosity regarding the man's extraordinary claims which would, he assured Tom, when he was ready to release them, change the face of Bigfoot hunting forever.

From time to time Tom visited the project, flying up from Texas and often bringing with him some very interesting people. Among them I recall Colonel Charles Askins, a US Army Champion Pistol Shot; Tom's brother-in-law, Lou Moorman, with his two sons; Nancy DeHerrera, a beautiful Beverly Hills socialite, who was one of his lovers in those days; Cathy McLean, from New York, a gorgeous, one-time model, another mistress; and Jeri Walsh, his social secretary, another paramour, a very well-endowed, cool, and clever lady who had previously, it was said, been a private secretary to Richard Nixon. And sometimes his own family came along, mainly his two sons, Tom Jr. and Charles. He brought them with him when he came to visit me towards the end of 1960, our first year with the project.

He called a couple of weeks in advance to say that he was coming, that he and the boys would like to camp somewhere in the Six Rivers National Forest and also that he would like to have as many of the project team join him in camp as possible. The group was also to include his private caterers, from Redding, a man called Red Matthes and his wife, whom I was to call without delay and book their services for his visit. This efficient and very pleasant couple were on call for all of Tom's camping visits ever since his first visit, when he'd had to suffer Peter Byrne's very questionable camp cooking.

He also gave me, when he called, some interesting news. On this visit, he said, the elusive taxidermist would be coming to join us, to reveal, at long last, his secret and extraordinary project and to bring it to a conclusion—which, Tom said to me in confidence, after two years of waiting and a large amount of money spent on his honorariums and expenses, was about time.

Tom flew into Redding, bringing with him his two sons. There they rented a car and drove up to Willow Creek, where they put up for the night in a motel. The boys, tired after their journey from San Antonio, went to bed early and Tom and I had a drink and a snack at the local pub and a talk to catch up on events and planning. He also had a question for me but explained that before asking it he would have to tell me—even though he was not supposed to—about the taxidermist and his findings, because the man in question would be coming to join us at our campsite the following day and, if all went well, to reveal to the world, well, at least to Tom and possibly also to me, the extraordinary discovery he claimed to have made.

The man, Tom told me, claimed to have found and been studying nothing less than a Bigfoot defecation site, at a place where one or more of the creatures came to defecate as often as two or three times a week. He had found it, he said, by studying the movements of deer and bear and other animals in the general area of the site and by noting how they all seemed to avoid it. Their droppings, particularly those of the bears, could be found almost everywhere in the forest but never at this site, or anywhere near it.

This brought up Tom's question to me. Did I think that it was possible that a wild animal—which is how, he said, until we knew more, a Bigfoot could be classified—would do something like this? Chose a single place as a toilet and use it continuously? Animals don't practice hygiene, he said, adding, with a smile, the ever-amusing campfire joke…where does a bear poop in the woods, the answer to which is, of course, anywhere it wants to.

I thought about Tom's concern and realized that I had an answer, or maybe something close to an answer, one from my background in Asia. I told him that in my experience as a professional hunter and naturalist in the forests of India and Nepal, there actually was one animal which had this very unusual toiletry practice. This was the Indian one-horned rhino, an animal that chooses a single place for its daily defecation and uses it on a continuous basis for considerable lengths of time, eventually building a pile of dung that may be as much as four feet high. The animal returns to the same site every day, usually in the early evening and after a cautious approach, to make sure there are no lurking hazards—like tigers or poachers—then turns around and backs up to the pile. It then defecates and urinates and it continues to do this, at the same pile, until the mound of dung becomes too high and is physically inconvenient. When this happens, it abandons the pile and chooses another place and starts another pile. So, I explained to Tom, although there is a vast difference in species between an Asian rhino and a Pacific Northwest primate, a similarity of toiletry habits, while improbable, was possible.

Tom was encouraged by what I told him and we then split up, he going off to bed at the motel and I driving back to Salyer. Early the next morning, I met him and his sons with the Jeep and after a quick breakfast at the local café and a visit to the market for supplies, drove them the thirty-odd miles to a campsite in the Six Rivers Forest.

The campsite I had chosen for them was next to a creek called Bluff Creek, in the heart of the Six Rivers National Forest, and I chose it for two reasons. One, it was actually in an area that had produced some reports of Bigfoot activity—mainly footprints. And, two, it was close to where, as Tom had whispered to me the night before, our taxidermist claimed to have made his extraordinary find.

The name of the campsite I chose was Louse Camp and it is believed that the name dated back to a group of miners who camped there in the eighteen hundreds and who were probably not too particular about bathing, or personal hygiene, or too concerned about the mites and fleas and lice that lived with them, as their perennial companions, in their unwashed clothing. We presumed, when we camped there, as we did many times in the coming two years, that their unwelcome companions were long since gone.

In the afternoon, after we had set up our tents, Red Matthes and his wife drove into camp, hauling a huge utility trailer full of, as Tom's boys happily put it, all kinds of goodies, from bacon and eggs and biscuits and steaks to pop and candy and ice cream. Soon after Red and his wife arrived and when they were still setting up their professional kitchen, Steve Matthes appeared, driving his old pickup, with his little trailer coupled up behind. After him Bryan drove in, accompanied by Gerry Crew, and so by evening time there were nine of us in camp, all enjoying a fine, big campfire, a good dinner, and the joys of a well-stocked bar, something Tom always insisted on for his camps, he being a man who liked to play hard as well as work hard and, like me, enjoy life to the full. We all slept well that night, if only thanks to the subtle alchemist and a splendid dinner provided by Red Matthes and his diligent lady.

About midday next day we heard the sound of a car coming down the hill road above the camp and within minutes it drew into the campground and the man we were waiting for, the taxidermist, got out and walked into camp. I recognized him immediately because we had met once before, when, on Tom's instructions, I had stopped by for a quick visit—which was all I was to be allowed, with no questions to be asked—on my drive from Monterey to Willow Creek, one that took me through Redding, where the man lived.

He looked exactly the same as he had the year before: a little man, about five feet four, with a narrow face and hard eyes set rather too close together on either side of a sharp, beaked nose; above this was a shock of short, graying hair that stood straight up from the top of his head and looked as though it had been sheared off with a mowing machine. He was wearing shiny, store-bought black shoes, white socks, very stiff Levi trousers and a dark blue Levi jacket with metal buttons, worn over a white shirt. Even this outfit was exactly the same as on the day I had met him in Redding.

Now he greeted Tom, and, completely ignoring me and all of the others in the campsite, said to him, "Well, let's go," and, turning, walked back towards his car. I gave Tom a questioning look. But he shook his head and as he walked past me, whispering, "He doesn't want anyone else along." Saying which he climbed into the taxidermist's car, which took off up the road in a cloud of thick brown dust and quickly disappeared, leaving behind some puzzled looks and some obvious irritation at the ill-mannered way in which our little visitor had ignored all of the group.

What happened next, Tom said later, was the meat and potatoes of campfire stories, one which, while it left and will always leave a burning question behind, was probably best viewed as one of the more humorous of the many amusing stories that emerged from our Bigfoot hunting days of the sixties in northern California.

Tom and his companion drove up the logging spur road from the camp that joined the main Onion Mountain road, about a thousand feet above the campsite, and there turned right and headed north. After another mile they came to a halt. Tom was told, is a whisper, he said,

"This is it, this is where we go in," and he was warned that above everything else he must be absolutely quiet, with nothing but subdued hand gestures for communication between he and his guide.

Then, the little man in the lead, they plunged into the thick manzanita brush that is the principal foliage of the Onion Mountain area and made their way west, up a long, steep hillside, to where the hill topped out in a small clearing. The walk, or rather the slow, tiring climb through the thick brush, took close to thirty minutes. For the last thirty yards before they reached the clearing, Tom was told to crawl on his and knees, which he did, arriving at the edge of the clearing in a bath of sweat. He was thankful when his companion indicated that they had arrived at the site of his discovery and that they should then both now lie down in the brush at the edge of the clearing with nothing more than their heads showing.

The clearing was roughly circular and about thirty feet in diameter, covered with short grass and, with the exception of a narrow opening down at the far end, opposite to where the two men were now lying, walled in by thick, dark red manzanita brush. At the far end of the clearing, close to the opening, was a small tree and underneath it, sure enough, was a mound of beige-colored dung, close to two feet in height. Tom's little companion pointed to the mound and nodded. He was silent but his gesture said it all…there it is, this is what I have been telling you about, this is what we've been waiting for, all this time…a Bigfoot defecation site.

Tom said that he found the situation very exciting. If this was a Bigfoot feces drop, indicating that one or more of the creatures came here on a regular schedule, then the possibilities were enormous. He could erect watch towers in the area, bring in a helicopter, set up remote control cameras, invite scientists to join the team and, the most exciting thought of all, look, at long last, at the possibility of encountering and communicating with the one of the creatures. His thoughts, he told us, were racing through all kinds of scenarios when suddenly he heard the sound of breaking brush. It came from the far end of the clearing, where the gap was in the wall of manzanita brush, and when it quickly developed into what sounded like heavy, thumping footsteps, Tom said he felt the hair on the back of his neck stiffening at the thought of a giant primate of unknown physical potential suddenly stepping into the clearing, no more than twenty feet in front of him.

A little nervous and not quite sure of what he should do now, he was actually considering sliding back into the brush for better concealment. But what he saw then, emerging out of the brush through the little opening at the end of the clearing, was something different from what he was envisioning. It was an elderly man—obviously a Native American from his appearance—riding astride a small white pony which had wicker baskets strapped on either side of its leather saddle. He rode quietly into the clearing, slid off the pony, tied the animal to the little tree and then, his peripheral vision detecting the shapes of the two prone forms at the edge of the brush, turned to look directly at them and greet them with a friendly, "Hullo, gentlemen, can I help you?"

Tom, slightly embarrassed at the somewhat ludicrous situation he suddenly found himself in, climbed to his feet, walked up to the elderly man and put out his hand to greet him and ask him, very politely, what he was doing there. The old man might have said, "I'd like to know the same of you two." But he replied, shaking Tom's hand gently, that he had come to gather herbs, that he had been doing so for years, that he always came to this same place and, pointing, always tied his pony to the same old tree. At which point the pony gave a healthy fart and with a couple of gentle popping sounds, added another set of droppings to the pile.

As Tom said, the story was one that would contribute to the humor of many a campfire evening to come. The only question that it left, however, was, did the little taxidermist really think that what he said he had discovered and was supposedly studying was Bigfoot dung? Or did he know that it was not, with the whole escapade planned as nothing more than a clever scam, one that until then had provided him with two years of regular salary and generous expenses? We shall probably never know. Returning to camp with Tom, the little fellow stayed in his car out on the road and did not come into the campsite. When Tom stepped out he reached over, slammed the door, gunned the engine and took off in a cloud of dust. We never saw him again.

My sixties Bigfoot project attracted several other colorful characters like our little taxidermist. But their scamming attempts were more laughable than anything else. One was a man who claimed to have a young Bigfoot in captivity and wanted to sell it. He persuaded me to let him meet with Tom on one of his visits, to discuss a sale, and one sunny morning at the café in Willow Creek he joined us. He had one in captivity, a young one, he said, maybe a teenager, about 350 lbs in weight, and he would sell it for $25,000. He kept it in a cage behind his house and it was eating so much food—Frosted Flakes, I think he said fifty pounds a day, and costing him so much to keep that he had to get rid of it. Without hesitation, Tom wrote out a check to the man for $25,000.00 and put it on the table in front of him. But when the man reached for it, Tom drew it gently back and said, "It's all yours. It's made out to you. But, er, if you don't mind, we'd just like to have a look at the creature before you take the check."

Mumbling, "Well, I can probably arrange that. Let me just go home and talk to the missus. I'll meet you here tomorrow," the man left.

We met the man next day, by appointment, same time, same place and his was a tale of woe. The poor little thing—which is what he called his captive Bigfoot—had gotten sick in the night and, scared that it might die on him, he had opened the cage and let it go.

There were one or two others like this one, some of them jokers, some of them outright hoaxers. But their stories have little place in this narrative and, compared to the decades of sincere and dedicated work of investigation and research done by so many decent people across the years, they are best consigned to oblivion.

I do not know how long my project might have lasted. As I mentioned earlier, while the Himalayan searches had a definite time frame—three years—our Pacific Northwest project had none. The original project, with the 1959 group, was open ended. Its duration was to be based on the field work and reporting of the '59 group and when this failed to materialize, Tom simply brought it to an end. My Pacific Northwest project with Tom might have gone on for quite a few more years because he was happy with my group and our work and also because he had the interest, and the funds, to keep it going. And though I often thought about going back to big game hunting, had this Bigfoot work not come to an end quite unexpectedly, I believe that I might have stayed with it. I liked the work, enjoyed the companionship of my team and working with Tom and his associates, and just loved—and still do—the great forests and rugged ranges of the Pacific Northwest. But, as so often happens, the fates decided otherwise and it all came to an abrupt end with the death of Tom in a disastrous plane crash in October, 1962.

He had gone to British Columbia to investigate some Bigfoot sighting reports and also to look at the possibility of opening up a second research project there which, he indicated, my brother Bryan might be interested in operating. Then, on October 6th, 1962, he boarded a small two-seater plane with one other man, the pilot, to fly back to the US. His route was one that would take him over the rugged country of eastern Montana. In the air, not far from the town of Dillon, the aircraft encountered bad weather. Now, in hindsight, I believe, as do several of my flying friends, that the pilot should have turned back. But apparently he decided to take a chance and continue. What happened then, to the little plane, may be what is nowadays known as wind shear. Whatever it was, it included powerful air turbulence, wind blasts violent enough to tear the right wing right off the aircraft and send the machine hurtling to earth. Tom and his pilot were killed the instant the plane hit the ground; the sheared-off wing was found a mile away.

Tom Slick's death was a tragedy for his family and his friends. It also brought closure, for many years, to serious research into the Bigfoot mystery and, of course, to our Pacific Northwest Bigfoot project as well. Tom was a great man, admired by all who worked for him and were associated with him, as well as loved by many women. Even now, after all these years, he is still poignantly missed by those of us who spent time with him, not the least of whom is this writer.

In the world of Bigfoot, and Bigfoot hunting, much water has flowed under many a bridge since 1962. A large number of people have appeared on the Bigfoot scene, to join in what in what Tom used to call the ultimate quest, some of them with scientific credentials, some of them as simple field men and women, others as writers, hunters, and naturalists. And, as seems to happen with a phenomenon as bizarre as the Bigfoot mystery, some of them part of a motley collection of colorful characters, among them jokers, clowns, comedians, pranksters, fakers, fools, skeptics, cynics, wits, and wags, people of unbelievable naiveté, and downright charlatans. I will name none of these latter here nor shall I give any time to their antics, the basic cause of which still seems to me to be nothing more than a childish need for attention. One thing I will say and that is that their claims and their behavior, much of which has received unwarranted publicity, have cast a dark and unwanted shadow over the good work of many sincere people, as well as the credibility of the phenomenon and the belief of many that it is real.

When the sixties Bigfoot project came to an end I closed down the Salyer office, disposed of the equipment—mostly as gifts to friends and project associates—and each of us then went his separate way. Gerry Crew, a widower till then, remarried and settled down in the Willow Creek area. There he spent his remaining years, until his death in the nineties. His nephew Jim also married and then went back to college; he presently lives in Washington State. Our project ladies, Shirley and Antje, left us—as I have already mentioned—to marry, and my brother Bryan moved to Lake Tahoe where he met and married a Scottish lady; he later moved to Carson City, Nevada, where he lives to this day.

Steve Matthes moved from Pasa Robles and settled down in Carlotta, California, and lived there until his untimely death in1995. His one son, Eddie Matthes, became a dentist and settled in Boise, Idaho, where I met him recently. Tom's children, Tom Jr., Charles, and Patti, moved from San Antonio to Atlanta, GA, where they presently reside.

The Odens, from whom I rented the Salyer house, lived on for many years and then they passed away. Betty Allen, our great historian, is also gone now, as are our stalwart field volunteer members, Bob and Ardis McClellend of Willow Creek, and Ernie and Dorothy Alameda, owners of the Oaks Café in Hoopa, a place where we had many a fine meal in their enjoyable company. In fact, to the best of my knowledge, all of the older people of that era are gone now, as are Tom's romantic and beautiful lovers, Nancy De Herrera, Cathy McLean, and Jeri Walsh. The old man with the scythe, the Grim Reaper, spares no one when he feels it is time to cut them down.

With the closing of the project, I went back to Nepal and reopened my safari company. I kept it going till 1968 when I closed it down and switched to wildlife conservation, to this end establishing, with two doctors and two attorneys, in Washington, DC, the International Wildlife Conservation Society Inc. Using IWCS, as we called it, and working with the government of Nepal, I took my original big game hunting concession in far west Nepal and converted it into a protected wildlife sanctuary which we named The White Grass Plains Wildlife Reserve. Since then and for many years, through IWCS and a sister organization, The American Himalayan Foundation, I have designed and executed fifteen conservation and wildlife preservation projects within the 200,000-acre reserve. Twenty-five years in all with just two breaks for some more Bigfoot hunting, one in 1971 for a second project in northern Washington, and another in 1992, for a third project in northern Oregon.

The second Bigfoot project, which I named The Bigfoot Research Project 11, was backed by former associates of Tom Slick: Robert (Fearless) P. McCulloch, of the McCulloch Chain Saw Company; C.V.

Wood, a Beverly Hills-based, semi-retired oil man from Texas; and Carroll Shelby, the famous racing driver.

The northern Washington project centered around the findings, which included 16 MM footage, of a bear hunter and guide and we based it in a rather bleak little settlement—three or four houses—called Evans, north of Colville, not far from the Canadian border and close to where the man lived. I ran the project and, much the same as with the Salyer operation, brought in vehicles—leased from International Harvester—and field gear and set up an office with a secretary, salaried field staff, Dennis Jensen and Don Byington, both of Marcus, WA, a local volunteer team, and the bear hunter guide in question.

The guide was our principal field man and much depended on the validity of his findings—mainly footprints and handprints—and on the authenticity of the film footage, which we spent time studying. The project commenced January 1st, 1971, with much enthusiasm on the part of my team and my sponsors. But when, after just three months, our studies of the footage proved, beyond a shadow of a doubt, that it was faked—from which we concluded that all of his other "findings" were also almost certainly bogus as well—its sponsors became disillusioned and, quite rightfully, withdrew their sponsorship and asked me to close down the project.

I stayed on for a couple of weeks at Evans, then returned the vehicles to International Harvester, closed the office, and let the staff go. I did not have to do anything with our "bar" hunter, as he called himself. Fearing the wrath of the locals at the exposure of his shenanigans—which included much lying and deceit, as hoaxing always will—he took off in the night and hightailed it south to California, where he had a home. We never saw him again.

With the closing of the Washington project, I moved down to northern Oregon and took up residence in a city called The Dalles. I went there mainly because there seemed to be some credible evidence of Bigfoot in the general area. This proved to be true, with one set of sightings as far back as 1910 and three in the late sixties and early seventies; of these latter, one was by three businessmen, one was by a school teacher and his wife, and one by a dentist and his wife. These are detailed in my 1975 book.

I wrote *The Search for Bigfoot* in 1975, when living in The Dalles. There, with the backing of Tim Dinsdale's associate, my Boston-based friend, Robert Rines, and his foundation, The Academy of Applied Science, I set up a Bigfoot exhibition close to a local restaurant called Spooky's. After a year, at the invitation of the Hood River Chamber of Commerce, I closed this and moved it west to Hood River and set it up at that pleasant little town's Information Center. I operated it there for a year and then closed it down and, looking for a permanent place to live, found and bought a beautiful, twenty-three-acre piece of property, near the settlement of Parkdale, on the East Fork of the Hood River, where I planted a one-thousand-bush organic blueberry orchard.

I lived there until the early nineties at which time, just as in the seventies with Tom Slick's associates, I was approached by a Peoria, Illinois-based industrialist named David Ransburg and asked if I would be interested in starting up and running another project. I said yes and in 1992 opened up The Bigfoot Research Project 111. I based it at Parkdale, and it was similar in many ways to both of my previous projects, although it lasted much longer, five years, concluding in 1997.

Among its findings were, one, via an $85,000 study, that the 1967 Bigfoot footage was very possibly genuine. Two, a dozen or so credible sightings in northern Oregon between the fifties and the seventies. And, three, a route that the creatures might have used—and might still be using—from Timothy Lake, south of Mt. Hood, via Ape Crossing—so named on early maps—on Highway 26 and then along the top of the huge ridge that encloses the Hood River Valley on its east side, across the Columbia River, into Washington, then up the equally enormous ridge that rises out of the river opposite the city of Hood River to run north, one that in the twenties was known as Whistling Ape Ridge and was so named on maps of that time, and onwards into Washington.

The Hunt For the Bigfoot

That was the last of my three professional Bigfoot projects, by which I mean privately sponsored, full-time 24/7 undertakings, with a fully equipped office, salaried staff, vehicles, access to reconnaissance aircraft and helicopters, all essential field gear, a fixed base and full coverage for all expenses. When it came to an end, I moved to Portland for a year and then to Los Angeles for a while, mainly to be with my new-found partner, Georgia-born Cathy Griffin. Recently, firmly together in a delightful relationship, we left Los Angeles and moved to the beautiful Oregon coast, where we presently live in an old fishing cabin built in 1920—the year my parents were married in England, how time flies—on a meandering mountain river, not far from the sea. Here my partner, working from home, pursues her career as a freelance corporate head hunter and career counselor and I write—fourteen books down, three to go—and, yes, work part-time with a local group in pursuit of the elusive creatures of my dreams, still—though pushing ninety now—getting out and camping and searching, rain or shine, in the rugged Oregon coastal terrain and till hoping that one can be found before the grim old man with his wicked instrument comes for me.

40

The author, 1960, (age 35,) at Tenmile, southern Oregon, investigating a Bigfoot sighting by two boys who said they encountered a Sasquatch in open country, coming down a hill towards them. When the creature saw them, it simply changed course and kept going. The boy's parents did not believe them, which the boys found distressing. But the author did - finding their detailed description of what they saw to be very credible. Unbenownst to the boys, the very next day a logging truck driver reported another credible sighting, just two miles away from the location of their incident.

*Giant Bigfoot with female partner and juvenile, photographed by the author deep in the forests of the Pacific Northwest? Or maybe models for the author's 1978 documentary by Alan Landsburg Productions, of Los Angeles, CA, **MANBEAST! MYTH OR MONSTER**, directed by Nick Webster, (1912-2006) with cine photographer Bill Hartigan and shot on location in the Nepal Himalaya, London, England, Loch Ness, Scotland and the Pacific Northwest.*

Photo taken in 1973 at the author's Bigfoot observation post on Table Mountain in The Dalles, Oregon. Left to right, Tim Dinsdale, (1924-1987), of Redding, UK, the famous British Loch Ness monster hunter and author. Robert Rines (1922-2009), President of the Academy of Applied Science, Boston, MA. And Dennis Jensen (1942-2015), the author's Bigfoot hunting friend and partner of the seventies. These three men played an important part in the author's career in Bigfoot hunting, contributed much to his research and were lifelong friends.

*Left, veteran Bigfoot enthusiast and friend of the author of many years, John Cordell. Right, Hal Halderman, of Pacific City, OR, Bigfoot field research associate and long time colleague of the author. Present day Bigfoot research on the part of the author his associates concentrates-in addition to inquiries into eyewitness reports, historical data and field expeditions-mainly in the area of Motion Sensor cameras. In the absence of a major project of the types conducted by the author in the 60s and 90s, and with limited funding and time availability, the use of present day technology seems the best way to go. Using available instruments and situating them in locations determined by Geo Time Patterns- i.e., place and time equations based upon credible eye witness reports in the geographical locations of the incidents in question-sums up the present day work of the author and his small, select group of companions. The camera used is the **Bushnell Trophy Cam**, a first-rate model with excellent capability, extensive range and fast shutter speeds. Using this, while we have not yet succeeded in obtaining a photo of a Sasquatch, we have amassed an interesting collection of photographs of mountain lion, bear, deer and elk. (See photo next page.)*

A large Black bear photographed with one of the Halderman-Byrne team's **Bushnell Trophy Cam** *motion sensor cameras. The animal, standing up, reached a height of close to 7 feet and its estimated weight was about 375 lbs. Timing on the camera showed that it had been at the site about two hours before Hal and the author arrived for their weekly camera check. Usually harmless, the Black should always be regarded as potentially dangerous. While normally shy and even timid, as with many large wild animals there are circumstances under which it can be provoked and these include when it is feeding, when mating, when surprised by the unexpected appearance of a person, if and when it feels it is cornered, when injured and when, as a female, it has cubs. In Asia-an area of the author's background in wildlife studies-more people are savagely injured and killed by bears than by any other animal, including tiger and leopard.*

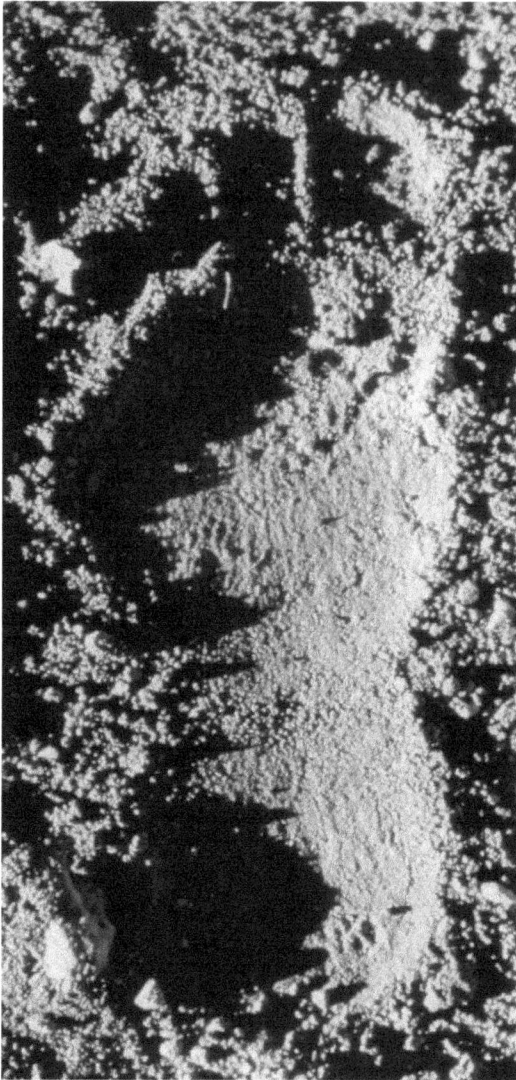

A fourteen and a half inch Bigfoot footprint found by the author and his 1960s Bigfoot Project team on the Bluff Creek road in the Six Rivers National Forest, northern California, summer, 1960. Note the bulge behind the big toe, something not seen in the human foot. This was one of about 300 prints found on this occasion. In all, during the thirty-four months of the author's 1960s Bigfoot project, four sets of Bigfoot prints, of similar or different sizes, were found.

Tom Slick and his sons, Tom Jr and Charles - who is looking towards camera - lunching at the author's Bigfoot project field base, Louse Camp, in the Bluff Creek watershed. Their companion is Tom's Redding-based camp caterer, Red Matthes. (See text and notes on the author's doubtful camp cooking expertise.) Tom and his fine family were frequent visitors to the Bigfoot Project and enjoyed hiking and camping-and Bigfoot hunting-with the author in the great forests of the Pacific Northwest.

Murray Field airport, located between Eureka and Arcata, on the coast, in northern California, to where it is believed Patterson and Gimlin drove, from Bluff Creek, after shooting the film footage and from where they air-freighted the famous Bigfoot footage, depositing it there on Friday, October 20th, 1967, for dispatch to Yakima, WA, on Saturday, October 21st. Much research has been done on this gray and puzzling area of the film footage... how and when it got from northern California to Yakima, WA, and precisely where and when it was processed and by whom. To date no one has been able to provide a satisfactory answer to these nagging questions.

The Search for Bigfoot 1975

FOREWORD, by Robert Rines: The intrigue of a search for an unknown species of man or animal, be it in the northwest wild s of the American continent or in the peaks of the Himalayas, sparks the souls and imagination of most people.

Couple such a concept with the relatively recent discovery of the coelacanth (a previously "extinct" prehistoric fish), and the much more recent discoveries of large, apparently unknown marine animals in northern Scotland 's Loch Ness and of an unknown tribe of primitive people in the Philippines and minds begin to open more freely to the possibility that the long list of prior accounts of the American Indians and settlers and more current reported sightings by apparently wholesome citizens, bolstered by a piece of amazing film footage and a myriad of footprint records, might be heralding another fascinating discovery in our own time, country, and continent!

It was in this vein, at the urging of Tim Dinsdale, my dear friend and companion in the continuing quest for definitive pictures at Loch Ness, that both Tim and I, a couple of winters ago, journeyed to The Dalles on Oregon's Columbia River, to see for ourselves what the intrepid explorer, Peter Byrne, was actually up to in his search for Bigfoot or Sasquatch.

After trekking with Peter through the snows in the forests of Mount Hood and neighboring mountains (and with me nearly disappearing up to my neck in a snow blanket that was deceptively covering a deep crevasse), we came away entirely convinced that there were hundreds of square miles of wilds, perhaps never frequented by man, that could harbor a missing tribe, and that patently contained more than adequate food and shelter for same.

We tried to analogize between Peter's "needle-in-a-hay-stack'" problems and our not-so-different problems at the Ness in determining where to be and at what time, in order-hopefully-to detect what we are respectively seeking. Common aspects of habitual rounds of the domain, food-seeking specified times of the year, luring to exposure, and sophisticatedly photographing to document, were hashed over in great detail around the campfire, in the dripping forests of the giant Douglas fir.

Convinced of Peter Byrne's sincerity and dedication, and thoroughly delighted with Peter's extreme expertise in animal tracking and related skills, Tim and I determined to provide what support we could to the Byrne expedition.

And to this end, through the efforts of the Academy of Applied Science in Boston, of which we are all members, we have been able to see to some funding, some photographic equipment, some night equipment and personnel assistance, contacts with the scientific community that could be available for assisting in study of the live creature, and now help in establishing a portable educational museum, currently in The Dalles, that serves as an information-receiving headquarters with a continuing educational display.

The historical and updated accounts of the exciting search for the Bigfoot or Sasquatch that Peter Byrne here provides in his own inimitable and refreshing writing style, should delight the reader, and bring to the Byrne camp an ever-increasing army of well-wishers and believers, so necessary for the lonely pioneer who must defy convention and the mainly unimaginative and usually self-a appointed custodians of our current scientific "knowledge." *Robert H. Rines, President, The Academy of Applied Science, Boston, Massachusetts.*

Preface to the 1975 Edition

Perhaps the reason why millions of people now find the Bigfoot legend fascinating could be that it is, in this day and age, one of the last great unsolved and unexplained mysteries of this shrinking world of ours. People find excitement in reading about it, in hearing about it and of course in talking about it with other people who have had some experience with the phenomenon, who perhaps have seen a footprint or even, in some cases, have seen one of the giant primates. But if people find this vicarious association with the phenomenon exciting, how much greater the thrill if they could actually take part in the searches that are presently being conducted in the Pacific Northwest, actually go into the mountains and personally hunt the elusive giants? I am fortunate in being one of the few people who has hunted for the mysterious giant hominids of America's Northwest and to date the only man alive who has made a profession out of this extraordinary search and who, through the support of many dedicated associates and sponsors, continues that profession on a full time basis for twelve months of every year.

People have asked me, what is it that keeps you going, now, after four years, in a search that may go on for another decade? What spurs you on, keeps you out in the wild lonely mountains and binds you to a life in the backwoods of America's Pacific Northwest? What is the driving force that propels relentlessly forward in a search that is not just for a needle in a haystack but for a moving needle that obviously does not want to be found?

The answer is difficult to put into words. Perhaps, with me, the reason is a simple one. Professionally and emotionally I am a hunter. Most of my life-some twenty years-has been spent as a professional big game hunter. I have hunted big game in both Africa and Asia, with a gun and, in recent years, after I gave up shooting, with a camera. (My professional hunting career ceased in 1968.) And to me, the title that I give to the search for the Bigfoot is one that I feel sums up its meaning most adequately. I call it the Ultimate Hunt-a hunt that is so difficult and so demanding that none but the most tenacious of men-or, as some of my friends put it, the craziest of men-would dare to follow it. The quarry in this hunt-the trophy, if one wishes to call it that-is nothing more than the rarest of all game, a possibly highly intelligent, highly mobile, totally nomadic, partially nocturnal creature with a habitat that measures more than 100,000 square miles and that embraces some of the most difficult and dangerous country in the world.

 For sheer diversity there is nothing to equal the different types of terrain that this enormous area encompasses. The jungles of north India and southern Nepal, where I once hunted tiger and leopard, are dense and thick. But they are nothing compared to the near impenetrable scrub of the coast of British Columbia. The mountains of Kenya and the uplands of Mount Kilimanjaro have high rugged terrain where a lone hunter must watch the weather- as well as the mountain buffalo, rhino and elephant-if he is going to survive. But every year the highlands of Oregon and Washington, the central Cascades from 4000 feet upwards, claim victims, people who get lost and die of exposure and whose bodies are sometimes never found. The Nepal and Sikkim Himalaya is a huge area of predominantly mountain terrain, and we well realized its size when we-I and my companions of the Yeti expeditions-hunted the elusive snowmen there in the forties and fifties. But in size it cannot be compared to the habitat of the Bigfoot. The coast country of British Columbia alone could contain all of the Nepal Himalaya and the Sikkim and Bhutan ranges also.

The ultimate hunt. Now, unlike my safaris of previous years, it is a hunt with a camera and not a rifle. But though it may seem like an impossible dream, to me it is the ultimate challenge and one to which I am now totally dedicated. Every time that the phone rings at the Bigfoot Information Center in The Dalles, Oregon, there is the excitement of knowing that this could be the call that will lead my partners and me to the quarry. Every time a new sighting is made and we race to the scene, there is the thrill of knowing that this time we may get there quickly enough to track and find one of the giant primates while it is still in the area. Every forest road that we drive, every mountain that we climb, every gorge that we explore, may have our impossible dream right there, waiting for us, the ultimate prize of the true hunter and one to which I certainly intend devoting many more years of my life.

Peter Byrne, The Dalles, Oregon, 1975.

CHAPTER ONE

A Look Back in History

History shows that often the loudest skeptics are those who know little or nothing about the subject in question. They have not studied it and will not do so, for the very reason that they do not believe in it. Nevertheless they are prepared to take the time to pronounce judgment on it. IVAN SANDERSON, author, scientist, at Columbia, New Jersey, December, 1970.

According to Native American folklore, there were once a large number of Sasquatch living on Vancouver Island, a large island, 12,408 square miles in area, off the south west coast of British Columbia. The Indians knew about them, feared them, and respected them, but granted that they were harmless. One of the Indians of the Nootka Tribe, who lived at Nootka in 1928, claims to have been carried off by them and held captive for some time.

The story, told to me by Father Anthony Terhaar of Mt. Angel Abbey in Oregon, is a curious one. Father Anthony, a much-loved missionary priest who traveled the west coast of Vancouver Island for many years, was living at Nootka at the time of the story and he knew Muchalat Harry very well.

Muchalat Harry was a trapper and something of a rarity among his fellow tribesmen. He was, according to Father Anthony, a tough, fearless man, of excellent physique. In the course of his trapping, he was wont to spend long weeks in the forest alone, something that the average Indian did not do in those days. The Indians of the coast were apparently a rather timid people and they seemed to regard the deep forest as the home and territory of the Bigfeet. When they went into the deep inland forest for any reason, they never went alone. Muchalat Harry was different from other Indians. He went in the forest alone and feared nothing.

Late one autumn Muchalat Harry set off for the woods with his traps and gun camping gear. His plan was to set out a trap line and stay in the woods for several months. He headed for his favorite hunting area, the Conuma River, at the head of Tlupana Inlet. From Nootka he paddled his own canoe to the mouth of the Conuma. There he cached the canoe and headed upstream on foot. Approximately twelve miles up-stream he made his base camp and, after building himself a lean-to, started to put out his trap line.

One night, while wrapped in his blankets and clad only in his underwear, he was suddenly picked up by a huge male Bigfoot and carried off into the hills. He was not carried very far, probably a distance of about two or three miles, at the most. When daylight came he was able to see that he was in a sort of camp, under a high rock shelf and surrounded by some twenty Bigfeet. They were of all sexes and sizes. For some time they stood around him and stared at him. The males were to the front of the curious group, females behind them and young ones to the rear.

Muchalat Harry was frightened at first and his fear grew to terror when he noticed, he said, the large number of bones lying around the campsite. When he saw these he was convinced that the Bigfeet were going to eat him. But the Bigfeet did not harm him in any way. Occasionally one came forward and touched him, as if feeling him, and when they discovered that his "skin" was loose – it was in fact his woolen underwear - several came up and pulled at it gently.

While they looked at him and examined him, Muchalat Harry sat with his back to the rock wall and did not move. He was cold and hungry, but his thoughts were only on escape. Sometime in the late afternoon, curiosity on the part of the Bigfeet seemed to slacken and with most of the creatures out of camp, probably food-gathering he thought, there came the opportunity that he needed. He leapt to his feet and ran for his life, never looking back. He ran downhill, toward where he guessed the river to be and sure enough, he soon came to his campsite. In what must have been blind panic he bypassed his camp and ran - *for twelve miles* - to where his canoe was cached at the mouth of the Conuma River.

The story of Muchalat Harry's arrival at the Nootka settlement is described by Father Anthony, as follows. It was probably three in the morning. He and his brother Benedictines were asleep and the village was quiet. Suddenly there was a series of wild cries from the waters of the inlet. Lights were lit and he and others hurried down to the water's edge. There, slumped in his canoe, near dead from hypothermia and thoroughly exhausted, they found Muchalat Harry. He was barefoot and clad only in his wet and torn underwear and he had paddled his canoe all through the winter night, 45 miles, from the mouth of the Conuma River. Father Anthony and his companions carried the almost lifeless form up from the water's edge. It took three weeks to nurse Muchalat Harry back to sanity and good health. The nursing was done by Father Anthony, who took him into his own care, and he told me that during the course of these three weeks, Muchalat Harry's hair turned to pure white.

The story of the kidnapping came out slowly. At first Muchalat Harry would talk to no one. Then he told Father Anthony what had happened and, later, others. When he was fully recovered to health he was asked when he planned to go back to collect his belongings, the camp equipment, his pots and pans, his trap line and above all, his rifle, at the lean-to on the Conuma. In 1928 a trap line and all of its pieces must have been worth a great deal to an island Indian. A rifle alone would be regarded as a highly prized possession. But Muchalat Harry never went back to the Conuma. Not only did he never return there, according to Father Anthony, he never left the settlement at Nootka and never went into the forests again for the rest of his life. He preferred to lose all of his valuable and probably hard-won possessions, including, incredibly, his rifle, rather than risk another encounter with the Bigfeet.

Late in 1972 I had occasion to visit Vancouver Island. I was on a routine investigating trip and when I found myself at Nanaimo, not too far by road from the west coast and the scene of Muchalat Harry's adventure, I drove there. I stopped in the town of Gold River and obtained from the very helpful Royal Canadian Mounted Police some maps and instructions on how to get to the Conuma River area. Nowadays there is a logging road that runs all the way down to the mouth of the river, and one Sunday morning, with the logging trucks out of the way, I drove there and made camp on the Conuma. I spent several days there, walking the river bed and exploring. I tried to make a rough determination of where Muchalat Harry might have had his lean-to and I found a place that offered a good campsite, twelve miles from the mouth of the river on the edge of a series of high bluffs. The salmon were running in the Conuma while I was there and all night long I could hear them splashing up the shallow waters of the river. In the morning Black bear worked the river, getting the spawning salmon that had become stranded ashore in the night or got tangled in the limbs of the many fallen trees that lay in the river bed. I counted six bears in several days. The country was generally wild and deserted and the actual mouth of the Conuma, where it flowed into the salt waters of the inlet, was one of the most beautiful places I have ever seen. Some of the forest close to the river had been logged off, but the logging operation had been relocated and while I was there all was quiet. The days began with morning mists on the river and then warmed to the clear crispness of perfect autumn weather. Evenings were cool and damp and nights bright with a starlight that provided almost enough light to read. I found no sign of Bigfoot on the Conuma, nor any sign of Muchalat Harry's trap line or lean-to. I hardly expected to find anything of the latter, after forty-odd years. But even though Muchalat Harry was long gone, the river and the forest remained unchanged. The splashing salmon, the cold, clear water of the Conuma, the moss-covered banks, the shallow pools in the forest that the Conuma drained, that were the breeding places of the salmon, the river birds, the plodding bears, the deep silent waters of the inlet, all were as they must have been forty years before, when Muchalat Harry cached his canoe and made his camp there.

For most people the history of the Bigfoot phenomenon begins within the last decade or so, with stories of footprints being found in northern California and other parts of the Pacific Northwest. In actual fact it is much older and, in the northwest, can quite definitely be traced back to the 1850s. It is mostly to be found-as I and my associates have traced it-in the form of old newspaper stories of encounters, by various people, with large, ape-like creatures that were very possibly Bigfeet. A few of the stories are, of course, wildly apocryphal, the creation of imagination on the part of some of the early settlers. Others are much more credible.

The oldest of them all and one that is definitely "borderline" is from the Norse Sagas and describes an encounter by Leif Erikson and his men, during their first landing in the New World, with creatures that were pictured as "horribly ugly, hairy, swarthy and with great black eyes."

In the literature we must include, of course, the legends of the Indians of the Pacific Northwest- legends that have been handed down from father to son for hundreds of years-from northern California to north central British Columbia. Among the legends are the stories of the Giant Men of Mt. Shasta, the Stwanitie, or Stickmen, of the Washington mountains, and the Sasquatch, or giants, of the Salish Indian tribes of British Columbia.

Let us have a look at some of these stories in the light of their being "Bigfoot" evidence, starting at the earliest-Leif Erikson's strange encounter-and following them through to the present day. Did the bold Leif and his Norsemen encounter Bigfoot? It is possible. It is more probable that the creatures that they encountered were simply Indians. The Norse word "*skellring*" is a term of contempt. It means, roughly, a "barbarian." But what caught my eye when reading Samuel Eliot Morison's account of the early Norse voyages, was the word "hairy." The Norse were a hairy people themselves, big men with no doubt, matted hair and beards. Why did they remark on the "*skellring*" being hairy? Was it because they were very much hairier than the Norsemen, even covered with hair, perhaps? If the encounter had been between, say, Tibetans, who are not a hirsute people, and the "skellring," one could understand the reference to hairiness. But why the Norse mention? After a thousand years the mists of time have drawn their veil over the Norsemen and their heroic voyages into the New World, leaving us with an age-old question which we may never solve.

Leaving the Norsemen and taking a huge leap forward in time, we come to 1810 and the overland journeys of a Northwest Fur and Trading Company agent. His name was David Thompson and he was, as his diaries show, a bright young man who took an interest in what went on around him. His diaries also show him to be much traveled, and at one time when he passed down the great bend of the Columbia River where The Dalles is situated today, on one of his journeys he found what might well have been Bigfoot tracks. At the time of the footprint finding reported in his diaries, he was making his way across the Rocky Mountains at the place where Jasper lies today. The year was 1810 and the month was January. He was with a small party of Indian guides when he found the tracks. The prints were fourteen inches in length and some eight inches wide and they resembled, he wrote, the paw print of a big bear. But the nails seemed to be too short to be those of a bear and when Thompson talked with his Indian guides they told him that the prints were those of a mammoth-Thompson's word. He found another set of the same footprints in the fall of that year and he marveled at them, at their size and the Indians' telling him they were of a mammoth.

What made the footprints that he found? Was it an old grizzly bear with its claws worn down? Or a big Brown? The size of the prints were not out of proportion with those of one of the big bears. But if it was a grizzly would not his guides, Indians of the region, have recognized it as such? One wonders what Indian word it was that Thompson translated into mammoth. The Indians, old though their legends were, would hardly know of the Mammoth that once roamed the region, any more than they would have known of the modern elephant. That Thompson was intrigued by the tracks there is no doubt, for many years later he expanded his diaries into a narrative and in these later writings we still find him wondering what could have made those huge prints.

From 1810 we leap across another thirty years to 1840. From that year emerges a fascinating letter from a missionary living in northern Washington. We are indebted to Robert Ruby, M.D., of Moses Lake, Washington, for finding this letter and to the Holland Library of Washington State University for permission to publish it. Of particular interest in the letter-apart from the mention of "giants" that could well have been Bigfeet-are the notations concerning their strength ("they can carry two or three beams upon their back at once"), the size of their feet ("they say their track is about a foot and a half long") and their smell. As readers will know, many people who have been close to a Bigfoot say they have noticed the very strong smell that they seem to emit. Both Patterson and Gimlin, who made the 1967 Bigfoot footage, noticed this powerful odor and think that it was this that caused their horses to bolt. As to the stone throwing, we have several mentions of this in our records as well as the famous story from Ape Canyon of rock throwing.

To: The Rev. David Green. Date, April, 1840.

Secretary of the American Board of Commissioners for Foreign Missions.

Dear Sir:

I suppose that it is not necessary for me to say much as Mr. Eeles has given you all that is interesting. My health is very poor and I fear I shall be unable to pursue my labors as a missionary. We are, we hope, doing something in the language. It is, if I may be judge, very difficult and will require much hard study before we have much knowledge and are prepared to make any direct or forcible appeal to them. I should not be at all surprised if this mission prove a total failure. How much more confidence I should have in its success if we had had real opposition to encounter from the Indians at its commencement. We, fear, are destined to experience some opposition from the old chief I named in my letter last fall. He has been absent most of the time since. He soon after went off to Buffalo (hunt) and has not yet to my knowledge returned. I left our place last Monday for this place (Fort Colville) and shall leave tomorrow for home, "Deo volente" which I expect to reach in two days.

It has been very sickly in this region the last part of the winter. Many have died. I do not know what can be done to save them from utter extinction. They seem as fated to fade away before the whites as the game of their country. There seems but one way that they can be saved and that is by settling them and civilizing them and this I fear they cannot bear. I sometimes think that it will be as injurious to them as their superstitions which are carrying them off very fast. Whatever is done for them must be quickly done, for there will soon be nothing to labor for. We need to be placed in such a situation that we can devote all our time and energies to them and when that is done we can do little on account of the few that we have access to. I think that I may safely say that the two tribes, Nez Perces and Flat Heads are as well supplied with ministers as New England, that is, there are as many preachers compared to the number of the people. We can only have access to a few at a time. If we travel and visit with them at their places we can have but little influence over them.

I suppose you will bear with me if I trouble you with a little of their superstition, which has recently come to my knowledge. They believe in the existence of a race of Giants which inhabit a certain mountain off to the west of us. This mountain is covered with perpetual snow. They inhabit its top. They may be classed with Goldsmith's nocturnal class and they cannot see in the daytime. They hunt and do all of their work in the night. They are men stealers. They come to the people's lodges in the night when the people are asleep and take them and put them under their skins and take them to their place of abode without even waking. When they wake in the morning they are wholly lost, not knowing in what direction their home is. The account that they give of these Giants will in some measure correspond with the Bible account of this race of beings. They say their track is about a foot and a half long. They will carry two or three beams upon their back at once. They frequently come in the night and steal their salmon from their nets and eat them raw. If the people are awake they always know when they are coming very near, by their strong smell, which is most intolerable. It is not uncommon for them to come in the night and give three whistles and then the stones will begin to hit their houses. The people believe that they are still troubled with their nocturnal visits.

We need the prayers of the Church at home.

I am, My Dear Sir, Yours most truly and submissively, E. Walker (Elkanah Walker) Missionary to Spokane Indians.

In 1851 we find an account in the May 16 issue of the Times-Picayune, a New Orleans newspaper. The account appears to have been obtained from the Memphis Enquirer. It is also reported as having appeared in the Galveston Weekly Journal, of Galveston, Texas. News seemed to have been shared, or perhaps simply pirated, by newspapers in those days. Nowadays there are the wire services, UPI, AP, et cetera, but in the 1800s the only sources of news for many country papers, of events that occurred outside their own communities, were of ten other and larger newspapers. I was particularly interested in this story when I first saw it, for it is the only one occurring outside the Pacific Northwest that seems to be describing a Bigfoot. I spent many hours searching for the original Galveston Weekly journal, in the city library of Galveston. I was unable to find it. One wonders if the area of the sightings in Arkansas could have held an isolated community of the creatures, one that the Indians had driven into some secluded pockets of forest. I do not know how extensive were the forests of Arkansas in those days, but they must have been considerable before the arrival and advance of the main bodies of pioneers and settlers with their axes and saws. Even today there are, I understand, areas big enough to hold much game and each year people still get lost in them, sometimes for days before being found.

Sitting in the public library in Galveston and listening to the muddy waters of the Gulf of Mexico lap listlessly against the shore, I came across an item of the May, 1851, issue of the Weekly journal , an advertisement which offered "fast service to San Francisco and the West Coast." The service was via Cape Horn and the offer was for a "pleasantly quick ninety days to San Francisco." The writer, a born ad man no doubt, had either never traveled the Horn or else simply decided to ignore its rugged reputation. But thinking back on the distances involved and the slowness of travel in those days, I believe it unlikely that anyone in Galveston would have seen a news story in the July 4, 1884, issue of the Daily Colonist, a newspaper published in the port of Victoria, British Columbia.

The account, which is reproduced here, tells of the discovery of what could well have been a young Bigfoot, captured by railway men near the town of Yale, British Columbia-in an area where several sightings of strange creatures had been reported by line workers in previous weeks and held for a period by them. The "creature" was apparently found on a railway line by some of the workers and they captured it by, as the writer so delicately puts it, "taking a small piece of loose rock and letting it fall" (on the wretched creature's head). This had the desired effect of rendering poor Jacko (as they called the creature) "incapable of resistance for a time at least." In other words, they smashed the wretched thing on the head with a boulder and knocked it out.

The creature subsequently disappeared from the news. No one seems to know whether it died, or escaped, or was sold to a circus, as is rumored. My belief is that if the method of treatment while it was in captivity was in any way similar to the method of capture, it very quickly succumbed. In 1884, railway men were a rough and ready lot, and a quick exit to another part of this world, or to the next, was probably the best thing that could have happened to "poor Jacko."

From more than a thousand miles to the south and two years later comes the next story. It is contained in the January 2, 1886, issue of the Del Norte Record and is in the form of a letter from a correspondent in Happy Camp, California. The story, for which I am indebted to my friend George Haas of Oakland, California, is an interesting one.

Happy Camp is a small town in northern California and is not much changed since the time of the story. The Marble Mountains area- today the Marble Mountains Wilderness-is as wild and untamed as it was in 1886. In recent years I have trekked through it, caught fish in its lakes and camped in its silent forests. I have peered into many of its caves and into not a few of its old, abandoned mine shafts; I explored many of these long before I heard of the story. L never found anything there, no footprints in the caves or mine shafts nor any sign of a Bigfoot in the area. But that is not to say that there are no Bigfeet in the area now, or when Mr. Jack Dover, a "trustworthy citizen," claimed to have seen one in 1886.

Here is the Happy Camp correspondent's letter dated January 2nd. 1886.

I do not remember to have seen any reference to the "Wild Man" which haunts this part of the country, so I shall allude to him briefly. Not a great while since, Mr. Jack Dover, one of our most trustworthy citizens, while hunting saw an object standing one hundred and fifty yards from him picking berries or tender shoots from the bushes. The thing was of gigantic size-about seven feet high-with a bull dog head, short ears and long hair; it was also furnished with a beard, and was free from hair on such parts of its body as is common among men. Its voice was shrill, or soprano, and very human, like that of a woman in great fear. Mr. Dover could not see its foot-prints as it walked on hard soil. He aimed his gun at the animal, or whatever it is, several times, but because it was so human would not shoot. The range of the curiosity is between Marble mountain and the vicinity of Happy Camp. A number of people have seen it and all agree in their descriptions except some make it taller than others. It is apparently herbivorous and makes winter quarters in some of the caves of the Marble Mountains.

In 1891, Theodore Roosevelt traveled through Yellowstone Park and heard various stories of Bigfoot sightings. One of these he retold in THE WILDERNESS HUNTER, published in 1893. Roosevelt calls it a ghost story, and recounts it as told by a "grizzled, weather-beaten old mountain hunter, named Bauman, who was born and had passed all his life on the frontier."

When the event occurred, Bauman was still a young man, and was trapping with a partner among the mountains dividing the forks of the Salmon from the head of Wisdom River. Not having had much luck, he and his partner determined to go up into a particularly wild and lonely pass through which ran a small stream said to contain many beaver. The pass had an evil reputation because the year before a solitary hunter who had wandered into it was there slain, seemingly by a wild beast, the half-eaten remains being afterwards found by some mining prospectors who had passed his camp only the night before.

The memory of this event, however, weighed very lightly with the two trappers, who were as adventurous and hardy as others of their kind. They took their two lean mountain ponies to the foot of the pass, where they left them in an open beaver meadow, the rocky timber-clad ground being from thence onwards impracticable for horses. They then struck out on foot through the vast, gloomy forest, and in about four hours reached a little open glade where they concluded to camp, as signs of game were plenty.

There was still an hour or two of daylight left, and after building a brush lean-to and throwing down the opening their packs, they started up stream. The country was very dense and hard to travel through, as there was much down timber, although here and there the somber woodland was broken by small glades of mountain grass.

At dusk they again reached camp. The glade in which it was pitched was not many yards wide, the tall, close-set pines and firs rising round it like a wall. On one side was a little stream, beyond which rose the steep mountain slopes, covered with the unbroken growth of the evergreen forest.

They were surprised to find that during their short absence something, apparently a bear, had visited camp, and had rummaged about among their things, scattering the contents of their packs, and in sheer wantonness destroying their lean-to. The footprints of the beast were quite plain, but at first they paid no particular heed to them, busying themselves with rebuilding the lean-to, laying out their beds and stores, and lighting the fire.

While Bauman was making ready supper, it being already dark, his companion began to examine the tracks more closely, and soon took a brand from the fire to follow them up, where the intruder had walked along a game trail after leaving the camp. When the brand flickered out, he returned and took another, repeating his inspection of the footprints very closely. Coming back to the fire, he stood by it a minute or two, peering out into the darkness, and suddenly remarked, "Bauman, that bear has been walking on two legs." Bauman laughed at this, but his partner insisted that he was right, and upon again examining the tracks with a torch, they certainly did seem to be made by but two paws, or feet. However, it was too dark to make sure. After discussing whether the footprints could possibly be those of a human being, and coming to the conclusion that they could not be, the two men rolled up in their blankets, and went to sleep under the lean-to.

At midnight Bauman was awakened by some noise, and sat up in his blankets. As he did so his nostrils were struck by a strong, wild-beast odor, and he caught the loom of a great body in the darkness at the mouth of the lean-to. Grasping his rifle, he fired at the vague, threatening shadow, but must have missed, for immediately afterwards he heard the smashing of the underwood as the thing, whatever it was, rushed off into the impenetrable blackness of the forest and the night.

After this the two men slept but little, sitting up by the rekindled fire, but they heard nothing more. In the morning they started out to look at the few traps they had set the previous evening and to put out new ones. By an unspoken agreement they kept together all day, and returned to camp towards evening.

On nearing it they saw, hardly to their astonishment, that the lean-to had been again torn down. The visitor of the preceding day had returned, and in wanton malice had tossed about their camp kit and bedding, and destroyed the shanty. The ground was marked up .by its tracks, and on leaving the camp it had gone along the soft earth by the brook, where the footprints were as plain as if on snow, and, after a careful scrutiny of the *trail, it certainly did seem as if, whatever the thing was, it had walked off on but two legs.*

The men, thoroughly uneasy, gathered a great heap of dead logs, and kept up a roaring fire throughout the night, one or the other sitting on guard most of the time. About midnight the thing came down through the forest opposite, across the brook, and stayed there on the hill-side for nearly an hour. They could hear the branches crackle as it moved about, and several times it uttered a harsh, grating, long-drawn moan, a peculiarly sinister sound. Yet it did not venture near the fire.

In the morning the two trappers, after discussing the strange events of the last thirty-six hours, decided that they would shoulder their packs and leave the valley that afternoon. They were the more ready to do this because in spite of seeing a good deal of game sign they had caught very little fur. However, it was necessary first to go along the line of their traps and gather them, and this they started out to do.

All the morning they kept together, picking up trap after trap, each one empty. On first leaving camp they had the disagreeable sensation of being followed. In the dense spruce thickets they occasionally heard a branch snap after they had passed; and now and then there were slight rustling noises among the small pines to one side of them.

At noon they were back within a couple of miles of camp. In the high, bright sunlight their fears seemed absurd to the two armed men, accustomed as they were, through long years of lonely wandering in the wilderness, to face every kind of danger from man, brute, or element. There were still three beaver traps to collect from a little pond in a wide ravine nearby. Bauman volunteered to gather these and bring them in, while his companion went ahead to camp and made ready the packs.

On reaching the pond Bauman found three beaver in the traps, one of which had been pulled loose and carried into a beaver house. He took several hours in securing and preparing the beaver, and when he started homewards he marked with some uneasiness how low the sun was getting.

As he hurried towards camp, under the tall trees, the silence and desolation of the forest weighed on him. His feet made no sound on the pine needles, and the slanting sun rays, striking through among the straight trunks, made a gray twilight in which objects at a distance glimmered indistinctly. There was nothing to break the ghostly stillness which, when there is no breeze, always broods over these somber primeval forests.

At last he came to the edge of the little glade where the camp lay, and shouted as he approached it, but got no answer. The camp fire had gone out, though the thin blue smoke was still curling upwards. Near it lay the packs, wrapped and arranged. At first Bauman could see nobody; nor did he receive an answer to his call. Stepping forward he again shouted, and as he did so his eye fell on the body of his friend, stretched beside the trunk of a great fallen spruce. Rushing towards it the horrified trapper found that the body was still warm, but that the neck was broken, while there were four great fang marks in the throat. The footprints of the unknown beast-creature, printed deep in the soft soil, told the whole story.

The unfortunate man, having finished his packing, had sat down on the spruce log with his face to the fire, and his back to the dense woods, to wait for his companion. While thus waiting, his monstrous assailant, which must have been lurking nearby in the woods, waiting for a chance to catch one of the adventurers unprepared, came silently up from behind, walking with long, noiseless steps, and seemingly still on two legs. Evidently unheard, it reached the man, and broke his neck by wrenching his head back with its forepaws, while it buried its teeth in his throat. It had not eaten the body, but apparently had romped and gamboled around it in uncouth, ferocious glee, occasionally rolling over and over it; and had then fled back into the soundless depths of the woods.

Bauman, utterly unnerved, and believing that the creature with which he had to deal was something either half human or half devil, some great goblin-beast, abandoned everything but his rifle and struck off at speed down the pass, not halting until he reached the beaver meadows where the hobbled ponies were still grazing. Mounting, he rode onwards through the night, until far beyond the reach of pursuit.

The Roosevelt story is interesting in that it is the only record-if indeed it is a record-of violence on the part of a Bigfoot. The injuries suffered by the man who died suggest a bear, rather than a Bigfoot. At the same time the footprints that Bauman saw were not, according to him, those of a quadruped, a bear.

Comes the new century and with it a spate of Bigfoot findings that begin in Coos County, in western Oregon's coast range country. A newspaper article in the Lane County Leader, April 7, 1904, talks of several incidents in that area. It mentions miners of that time and in quaint jargon tells of their sighting a "wild man" and adds, "They have seen him and know whereof they speak." I can find no other written reports from the Coos County area, but I have had several oral reports of "findings" in the last few years, the latest being in the summer of 1971. I spent some weeks in the area in 1972 and talked with some local people. There was nothing very definite in the way of new sightings or footprint findings, but the area, if one can believe all that one hears of it, does seem to have produced some credible evidence over the years.

In 1960 I went to Kelso, in the state of Washington, to see an elderly man named Fred Beck. Mr. Beck lived alone in a small cabin, was retired and was the only known survivor of what has become known as the Ape Canyon incident. The date of the incident was 1924 and Mr. Beck was the central figure in the incident. As the story went, he and three others had a mining claim in the canyon. On several occasions they found, he told me, large man-like footprints in the sand and gravel of the canyon bed. They wondered about what made them - but the pressure of work prevented them from worrying too much about the prints. Then, one day; one of the men saw what he thought was a large ape, peering at him over the top of a big rock. The man, Beck's companion, ran to the cabin that they had built close to the mine and got a rifle.

He fired a shot at the creature, which disappeared. A few days later, Beck himself saw one, walking ahead of him along a narrow trail high above the canyon. He was hunting at the time and carrying a rifle. He promptly shot the creature in the back. It turned on the trail and fell into the canyon. He did not see it again. That night and for several nights afterwards the cabin was the target of showers of rocks that fell on its roof and against its walls from the surrounding trees. Beck said that he and his companions rushed out with guns several times to see who or what was throwing the stones. They saw and heard nothing. As soon as they went inside the stoning started again and after a few nights of this, unnerved, they left the cabin and returned to Kelso.

Subsequently a group of men formed an expedition and went into the canyon to find the "giant apes" that Beck and his companions reported. They found nothing. The canyon, which appears to have been unnamed prior to the incident, was then given the name Ape Canyon and this is the name under which it is marked on all maps to this day.

Beck, showing me the rifle that he used to shoot the creature, told me that he never returned to the canyon and that the mining claim was abandoned. He did not know, when I talked with him in 1960, what had happened to the .other members of his 1924 party. He seemed to me to be honest and to be telling a true account of something that actually happened.

Early in 1961 I visited the canyon with two friends, Monte Bricker of Portland and Shearn Moody of Galveston, Texas. We planned to spend a week in the canyon and to search it from end to end. Unfortunately after only one day Monte had an accident. He scalded his foot in a pot of hot water on the campfire and had to be taken out to a hospital. Shearn Moody and I returned later and flew the canyon in a helicopter and still later I walked it out, from its mouth to its rugged, very steep end below an area called The Plains of Abraham. It is a wild, rugged, inhospitable area, and when I was told recently that it was "probably young boys, out for a lark," who were responsible for the stone-throwing in 1924, I could hardly suppress a smile. I was also informed, again recently, that it was well known that a local man had been responsible for the footprints in Ape Canyon in 1924. But what he was doing away out there in the wilderness, and who or what the creature was that Beck claims to have shot, is still unexplained.

In 1924, a group of miners working a mine on the east side of Mt. St. Helens returned to their homes in Longview with stories of being attacked by mysterious apemen. It seems that while in their cabin "huge creatures being at least seven feet tall and covered with long black hair" showered them with large boulders. The next morning they encountered the creatures and shot at them. One of the creatures was believed slain, but its body rolled over a cliff and into a deep ravine, destined to be known as thereafter as Ape Canyon.

Searching parties were immediately mobilized from the Longview area to visit the mountain region to seek out the headquarters of the apemen. These parties, however, found no trace of the apemen but did find the cabin with huge boulders around it and the inside torn to shreds.

Gorg Totsi, editor of the Red American, a weekly Hoquiam publication, published an explanation of the legend of the apes several years ago. He said these creatures were the ferocious Selahtik Indians, a tribe of renegade marauders, much like giant apes in appearance, living like animals in caves in the High Cascades.

Someone might, one day, come across the bones of Fred Beck's "mountain devil" as he called it. But it is doubtful, after all this time. The remains of the cabin are still there and were seen in 1972 by some "Bigfoot searchers." And somewhere close by is the old mine shaft. So the cabin was there, the mine was there, and I believe Fred Beck and his worthy companions were there. As to whether the Bigfeet were there and were responsible for the vengeful stone-throwing is a matter that must remain, for the time being, or at least until someone goes in there and finds and digs up the bones of Fred Beck's "mountain devil," a mystery.

CHAPTER TWO

A Look at the Present

The exact answer to the origin of the genus homo has yet to be found but I believe that it will be found in the relatively near future. RICHARD LEAKY, speaking at Foxcroft School, Middleburg, VA, March 1975.

There are many more fascinating accounts of Bigfoot encounters from the late 1800s and all the way through to the present day-too many to tell here. Some of them are told in detail in other chapters of this book. For others, perhaps a brief mention will serve. The following report, dated April, 1969, comes to us from Mr. and Mrs. Robert L. Behme of Magalia, Butte County, California:

You may be interested to know that my husband and I have seen what we believe to be a Bigfoot in Butte County, California. To our knowledge, nothing of this nature has been seen here before. On April 16th, (1969) about midnight, we were driving along the road from Paradise to Stirling City. The surrounding country is thickly wooded, well-watered and crisscrossed by deep canyons. As we drove around a long curve our headlights shone on what appeared to be a man in a fur suit, crossing the road. For one moment we had a front view as he turned toward the car, then walked into the darkness. Our impressions are that he was over six feet tall, completely covered with short, black hair which seemed to be flecked either with white hairy patches or mud. His face was white and hairless although the features appeared as a blur. The eyes did not glow in the light as would the eyes of an animal. The head was small and came to a peak at the top. He was heavily built with particularly heavy legs. He did not run, but shuffled away with a definite limp, once turning his head to look back at our car. The following morning I returned to the area to look but could find nothing. The ground near the road is rock, gravel and hard clay.

We have lived in this area for nearly ten years. My husband is a writer and photographer specializing in outdoor stories for such publications as Field and Stream and Sports Afield. I mention this with the hope that you will believe we are reasonable people, not given to hallucinations brought on by the novelty of a back-woods road at midnight. Naturally we have given a great deal of thought to what it could have been, other than Bigfoot. A bear is out. There are bears in Butte County, but all are smaller and do not cross highways on their hind legs-especially when one is apparently sore. The idea of a hoax occurred to us. Chico State College is about thirty miles away and this is the sort of involved trick that might appeal to college students, except at on a week-night, during a non-hunting or fishing season, at midnight, this is a very lonely road. The chances of a motorist passing until morning are slim. We even thought of the possibility of someone bent on robbery expecting that a motorist would stop at such an apparition. But again the lack of traffic makes this very unlikely. By Mrs. Robert L. Behme. Credit George Haas, Oakland, CA.

Two deputy sheriffs of The Dalles (OR) Oliver Potter and Harry Gilpin, told the author of Bigfoot encounters. Potter on the Wind River (WA) while fishing and Gilpin on Highway 84 near The Dalles, while on duty, cruising in a patrol car, at night.

A young man doing fire watch service at one of the fire towers in the Timothy Lake area in Oregon told us he saw one in 1970. It was early morning and the Bigfoot was walking up the road toward the tower. Suddenly it turned off the road and walked into the trees and disappeared.

Two young men, summer camping in an area near The Dalles, said they saw one just before dawn one morning. Small, falling stones awakened them and they looked up to see it standing on a bank above their heads. They had a shotgun, so they fired at it in fright and then ran for their lives. Apparently they missed, for no carcass was found.

A deputy sheriff, Verlin Herrington, of the Grays Harbor County Sheriff's Force, W.A., stated that he saw one walking along a forest road in 1971. He stopped his patrol car and watched it for a few minutes, before it turned off the road and walked into the trees.

A logging contractor, Glen Thomas, told of watching three of the creatures digging for rodents in a rock pile in the high mountains near Tarzan Spring, in the Clackamas River watershed, near Estacada, in Oregon. The trio consisted of a male, a female and a young one. The male did most of the work, lifting rocks weighing more than three-hundred pounds while digging. With Tom Page, of Mentor, OH., and Dennis Jenson, associates of the Bigfoot Information Center in The Dalles, Oregon, I visited the area of the incident in 1972. We saw the hole in question and counted many other holes in the high broken ridges above the Clackamas River. A brief study of the rodents of the area showed them to be marmots, which would hibernate in the winter, at the time the logger said he saw the two creatures.

Two fishermen said they saw a Bigfoot swimming across one of the small lakes near Priest Lake in northern Idaho. Its arms were underwater and so they were unable to see just what stroke it was using to propel itself through the water. It climbed out of the water, shook the water from its arms only and then disappeared into the lake grass. This was in 1970.

Four young people, traveling in a car to Portland, Oregon, early one morning in 1969, on Highway 84, said they saw one sitting on a high rocky bluff near The Dalles, Oregon. They stopped the car, got out, and watched it for four or five minutes. It was still sitting there when they left. The time was just after five in the morning.

The president of a heavy equipment company in Portland stated that he saw one while fishing in northern California in 1960. He was with a companion and they became separated. He heard noises in the brush and looked up to see the Bigfoot walking through some small pines about thirty yards away. Apparently it had seen him, for it was already walking away. The pines, young trees in a plantation, stood about six feet in average height and the head and shoulders of the creature showed clearly above them. Later, when he looked for his companion, he found that the man had returned to camp. His companion later told him that he had wished to continue fishing but had an uncanny feeling that he was being watched. He grew uneasy and had returned to camp alone.

Many of the reported sightings have been subjected to intensive investigation, by scientists and laymen associated with the Bigfoot Information Center in The Dalles or by independent groups of part-time searchers and investigators. Instead of coming to an end, as they might do in the face of diligent scrutiny if they were hoaxed, the incidents have continued. Each year has produced its quota of both sightings and footprint findings and to date the number of sightings has remained constant at an average of about four a year. The record shows five for 1971, three for 1972, four for 1973, four for 1974 and three for 1975.

The Bigfoot Information Center, the central office of the present search and investigation project in the Pacific Northwest, has now expanded its activities to include a Bigfoot Board Of Examiners, a group of trained and experienced Bigfoot investigators whose task it is to examine all reported incidents in the Pacific Northwest and then submit their findings to the Information Center. In addition the Center has increased its associate membership. Its loose-knit group of watchers and reporters living throughout the northwest now numbers over one hundred. For an indication of the rapidly expanding activity of the Center and its research team, both the full-time team working from The Dalles and the associate investigators, one has only to look at the number of reports handled by the Center through 1975.

The record number that were reported and investigated by the team does not necessarily indicate an increase in the actual number of incidents. Rather it indicates a public awareness of the work of the Center, something that has reached out to people previously unaware both of its existence and of the fact that there is now an established clearing-house where their stories, no matter how far-fetched they may sound, will be given serious and confidential processing.

In the year 1974, on learning about the Center, people who had been shy about their sightings or other findings, or who had been held back by the fear of ridicule, came forward and informed the Center researchers. In all cases their findings were investigated and even where the reports turned out to be negative, e.g., bear prints, or even large human prints, or, in the case of two sightings, large roadside tree stumps seen by car lights at night, the Center researchers made sure that all reports were given at least some attention and that the people who provided the information were properly thanked for coming forward as they did.

One of the first sightings in 1974 took place while winter snow still lay deep on the ground. Just west of The Dalles, the freeway, Highway 84, carrying traffic east and west between Portland and Pendleton, sweeps past a cluster of houses and trailers that is a settlement called Rowena. At four o'clock on a freezing morning in mid-January, Deputy Sheriff Harry Gilpin, of The Dalles Sheriff's Department, was concluding a routine patrol. He was heading east, towards The Dalles and driving at approximately sixty MPH as he approached the big highway sign that signals the exit to Rowena. While he was still about 150 yards from the sign, he saw, in the extreme limits of the patrol car headlights, something which he described as "very tall, probably about seven feet," close to the center of the highway and the steel center rail that separates the double lanes of the east-west traffic. It looked vaguely like a man and it was walking away from the center rail, to the north. Gilpin, who is skeptical about the existence of the Bigfeet, is not certain at all what it was that he saw. He will not state definitely that it was a Bigfoot. He admits that he was tired and that whatever it was, it was already moving out of the limit of his patrol car headlights when he saw it. But he is puzzled as to what a man would be doing on the freeway at that hour. A very tall man. On a very cold morning.

The next report that we received at the Center in that year came from Florence, a little fishing town west of Eugene on the Oregon coast. Late in the evening of March 8th, one of our younger associates, Mike Kuhn, of The Dalles, called to say that there had been a sighting at Florence that morning, by a schoolboy. He gave us details that included the schoolboy's home address and telephone number. The boy's name was Nick Wells, age nine and we called his mother and talked with her. She confirmed that her son claimed to have seen a Bigfoot that morning while on the way to school. She said that she honestly believed him to be telling the truth. She added that some of the other schoolboys, his companions, claimed to have found footprints.

I left at once, taking with me Celia Killeen, co-editor of the Bigfoot News and my assistant at the Center. The distance to Florence is about three-hundred miles and we drove most of the night, arriving there at eight-thirty the following morning, our trusty Scout International 4X4 loaded with camping gear and provisions for a week.

We found the Wells' home and interviewed Nick, the boy who said that he had seen the Bigfoot. It was a Saturday and, school being closed, he agreed to accompany us to the place where the incident occurred. Nick told us that the previous morning, at about 8:45, he had seen a large, brown, hairy figure walking in the heavy scrub that fringes the road near the school. He said that the creature stood about six feet in height and was moving slowly. When it saw him it stopped, growled at him, and then moved on. It did not chase him, as subsequent newspaper reports claimed, but simply walked into the deep brush. Nick told us that he took one look and ran for his life, and his schoolteachers later confirmed that he arrived at the school white-faced and panting. We searched the area Nick indicated, and found only shoe prints. It appeared that during the midday break from school, the previous day, some fifty or more boys had come down to the area to "see the Bigfoot." Any footprints that the creature might have made were long buried under an army of shoe impressions.

We camped in the area for three days and searched the sand dunes north and south of Florence and the mud flats on both sides of the Siuslaw River. All that we found was some of the densest brush we had ever seen, the penetration of which, in the inevitable and constant coastal rain, made the Florence incident a memorable one. Later, assessing the sighting, we were of the opinion that the boy told the truth about the sighting and that he very probably did see a Bigfoot. The Siuslaw River cuts through the coastal range, flowing into the ocean at Florence and the country north and south of the township, within the coastal ranges, is heavily timbered and generally devoid of human habitation. The three day investigation concluded, we returned to base.

About a month after the incident we received a note from Mrs. Wells thanking us for our visit and for a copy of our report, which we had mailed to her. She enclosed with her note a most interesting letter, from a Mr. C. E. Dixon, of Bremerton, Washington, to young Nick Wells. Mr. Dixon wrote that he had never seen a Bigfoot but that he had seen tracks in the same area, in 1905. The letter, written in a fine strong hand, said that in April of that year, the year of the Lewis and Clark Fair, he and a companion were in the vicinity of Florence, mainly to have a look at the ocean, which neither of them had ever seen. They had walked "twenty miles by trail, to Florence" and arriving there obtained the help of an old Indigenous man, Charlie, to get them across the mud flats. It was in the flats close to the present site of the town, that they found several sets of huge "humanlike" footprints. At that time they had never heard of Bigfoot and so they put them down as grizzly prints. What puzzled them was the size of the tracks. They had no measuring tape and so they cut a stick and later measured the stick. The tracks were eighteen inches in length and there were no claw marks. Mr. Dixon concluded his letter by telling Nick, "Even if your Bigfoot should turn out to be a grizzly, this is something to brag about, for the grizzly is supposed to be extinct in this country."

Twice during the summer of 1974 an Indian boy, night fishing for sturgeon in the Columbia River, claimed two sightings of a Bigfoot. Both occasions were in the very early morning, just after first light, and both took place just east of Stevenson, Washington, close to the first tunnel through which the Stevenson-White Salmon road runs. The Bigfoot, we were told, was seen standing in the shallows, up to its waist in water and motionless. On each occasion, when sighted, the creature had left the water and retreated into the bushes. Searches of the area by Information Center investigators produced no footprints. But on each occasion strong Columbia gorge winds had smoothed the dry summer shore sand and both searches took place more than a week after each claimed sighting. After the sighting we kept the area under periodic surveillance. No further sightings were been reported in that area into 1975.

The National Observer for Thursday, August 22, 1975, gave the author a birthday present with a report of an unusual sighting in British Columbia. The story ran as follows: "Wayne Jones still isn't sure what it was that he says he saw standing next to a building at his boys' camp on Harrison Lake up in British Columbia one night a few weeks ago. It looked somewhat human, it walked on two feet, it had a rounded head with large ears, it stood nearly eight feet tall, and its whole body, except face and hands, was covered with hair."

The story went on to say that the new sighting was another "tantalizing piece in a puzzle for [Peter] Byrne, who has been looking for the storied Bigfoot in the Pacific Northwest since 1970."

We heard of the incident at the Information Center three days after it happened. We probably would have heard of it sooner, but Wayne Jones, the young man who claimed the sighting, did not talk about it to reporters and asked his fellow directors at the camp to keep the matter quiet.

We were fortunate at that time in having one of the Center's more active researchers, Stuart Mutch, working on some other leads about one-hundred miles away toward the coast. We immediately contacted Mutch and asked him to fly into Camp Dunbar and investigate the incident.

Camp Dunbar is a British Columbia, government operated camp for emotionally disturbed children. It lies on a peninsula of land at the north end of Harrison Lake, a long narrow lake north and east of Vancouver in southern British Columbia. Access to the camp is by boat, via the lake, or by plane, to a narrow landing strip in the camp. The purpose of the camp is to provide quiet and peaceful surroundings, in a natural setting of woodland and water, as part of the rehabilitation program for children with emotional problems. Situated as it is in an isolated place, the camp has few visitors and the setting is one that is generally regarded by the government as well suited to its program.

Our associate, Mutch, flew into the camp in his own plane to meet with Wayne Jones, the camp director who claimed the sighting. He learned that Jones had been sitting by a campfire in the evening, about nine. Suddenly a Bigfoot had appeared out of the heavy forest behind his campfire. It walked slowly toward the fire and then stood and looked at him. Jones kept quite still. The creature, manlike in stance and in facial appearance, was, in view for perhaps three or four minutes and at a distance of about thirty-five feet it stood and watched Jones and the fire. It did not appear to be threatening. It made no aggressive movements. Its face, Jones said, was curiously human and not at all apelike. Suddenly, some of the camp's children came running through the trees. The Bigfoot turned and moved quickly toward the edge of the forest. It then walked into the trees and disappeared. Some of the children saw it. Later, some footprints were found.

Mutch flew into the camp a few days after the sighting and talked with one of the camp directors. Jones was not in camp at the time. When he landed a dozen or so of the children came up and surrounded his plane, talked to him and told him about the incident. But Mutch also noticed two men, and a little later a third, standing back along the edge of the airstrip, holding high-powered rifles and watching him. None of them approached and Mutch became a little apprehensive. Later he learned that they were there to hunt and kill the Bigfoot. They were Canadians, from Harrison Hot Springs and from Richmond. Why they were allowed into a camp of emotionally disturbed children, with loaded high-powered rifles, to hunt and kill something that had done no harm, had not threatened anyone, and had not even frightened any of the children, was not clear. Mutch stayed for a short time and then flew out. He did not see any footprints and thus the full value of the sighting is not determined at this time. However, Jones' story is generally believed by the other camp directors and his reputation as an employee of the British Columbia state government is believed to be sound.

Not far from The Dalles, Oregon, but in Washington state, north of the Columbia River, is the little township of Willard. Willard has a timber plant with one of the last sluice or water transporters in this country, down which the mill shoots its cut timber to a second plant, twenty-five miles away on the Columbia. Northeast of Willard is a small campground called the Oklahoma Campground. In mid-August a fisherman from White Salmon, a young man just a few weeks out of the army, drove up to Oklahoma Campground and then walked north a short distance to where a stream runs parallel to the camp-ground road. The campground itself is set in the heart of the rugged country that is the central Cascades, and north of it the road comes to an end. The young man planned to spend a day fishing and he picked an area with which he was familiar, that he had known since boyhood. He turned off the road and started to walk down to the stream. On the opposite side of the stream was another road that had been severed cleanly by the stream, leaving a jutting fragment of roadway that ended abruptly at the stream's bank. On this roadway, right on the edge, according to the fisherman, squatted a Bigfoot. The young man stopped, took one look and then turned and ran for his car. As he turned he saw the creature rising from the ground and starting to walk away. The frightened fisherman reached his car, jumped in, locked his doors, and drove wildly for Willard and the nearest telephone. From Willard he telephoned the Sheriff's Office in Stevenson. The deputy who took the call promised to investigate and later two deputies went to the area. They found nothing and a subsequent investigation by Information Center investigators also produced nothing. However, the young man's story had a ring of truth to it and it is indeed possible that it was a Bigfoot that he saw and that its reaction to him was exactly the same as his reaction to it, a fast exit into the deep forest of the mountains north of the campground.

The general pattern of Bigfoot sightings and footprint findings is centered in the coast ranges and the Cascade Range of the Pacific Northwest and the coast ranges of British Columbia. This is what is known to the cognoscenti of the Bigfoot fraternity as the First Area of Evidence, or Area I. For a hundred and fifty years this pattern has seldom changed. When it has, when evidence of Bigfoot activity in other areas has appeared, it has been restricted to what is called Area II and this restriction is one that is confining and definite. Area II includes only south central and southwestern British Columbia and one area of north central Idaho. There are no records in the files of the Bigfoot Information Center, either from sightings or from footprint findings or other evidence, of any credible Bigfoot activity outside of these areas. Within Area II evidence is sparse. Idaho, in spite of its wild and rugged mountains and wilderness areas, has produced no evidence for nearly ten years, and southern British Columbia has produced very little.

One of the few recent sightings reported to us at the Center from Area II took place in southwestern British Columbia in 1974. It was reported to us by a young couple who must, at their own request, remain anonymous but who personally came to the Information Center's Exhibition in The Dalles and told us the story.

They were driving at night, just north of Castlegar, in British Columbia, a town not far north of the Washington border. The weather was clear and visibility was good. As they came around a corner on the Castlegar-Silverton highway, they saw a huge, dark brown, or black, hairy figure standing on the edge of the hardtop. Both of them saw it at the same time and both were shocked at what they saw. They were adamant in their description that the creature was not a bear. They saw its arms, clearly, hanging by its side and they saw its head, well-rounded and not at all bearlike. The creature stood perfectly still as they passed. They did not stop. They did not turn around and go back. Theirs was an eerie feeling, seeing that giant lonely creature standing solitary on that bleak roadside. They felt, with a gentle philosophy which we admired, that perhaps it was best left alone.

The large-scale map that hangs in the Bigfoot Exhibition in The Dalles includes the states of Washington, Oregon and Idaho. Clearly marked, with red and yellow markers, are all of the sightings and footprint findings that the Information Center has recorded and investigated in the last four years. The markers range out of northern California and follow the mountain chains into southern Oregon. There they divide, as do the mountains themselves, soon after they pass the California border and split into two formations. A line drawn east-west through the city of Eugene marks the end of the main range which, in the upper drainage of the great Willamette River valley, splits in two, one arm becoming the coast range and the other the Cascade range. The lines of markers that indicate the range of the creatures also divide at this point. (Indeed, if they were to continue up and into the heavily populated valley of the Willamette, with its farms, road systems, cities and industrial areas, we would be very skeptical about them.) One line of markers is seen to continue north through the Cascades, staying close to the high backbone of the range, and passing on either side of mountains of considerable size, like Bachelor, the Three Sisters, Mt. Jefferson and Mt. Hood.

Crossing the northern border of Oregon they continue into the Washington Cascades, passing between and around Mt. Adams, Mt. Saint Helens, Mt. Rainier and Mt. Baker. From there they leave the United States and merge, northwest, into the coast ranges of British Columbia.

The second set of markers, the western set as they might be called, lie in the heart of the coast ranges and they dot the chain all the way through Oregon and into Washington, until they eventually wind into the Olympic Peninsula and come to an end hard against the San Juan Straits.

The Olympic Mountains have had their share of sightings and footprint findings, and the Exhibition map shows a dozen or so pins that mark incidents and findings reported to the Information Center over the years. One such sighting was reported in the Olympics late in 1974. Two young men from Port Angeles, Richard Taylor and Larry Followell, were doing some night driving on the Hurricane Ridge Road, not far from Port Angeles. Taylor thinks that he saw the creature first. Perhaps if Followell had seen it quicker he might not have crashed. As it happened, he saw it when there seemed no way to avoid it. It was close to the center of the road and it was a matter of swerve and avoid it or hit it head on. Followell swerved, left the road, hit a big rock and wrecked his car. Both men finished up in the nearby Olympic Memorial Hospital with injuries and both swore statements of the sighting to local police. Olympic Park Rangers went to the scene of the incident to search for any sign of the creature the men said they had seen. No footprints were found, but this was not surprising, for the ground was frozen hard. The month was December-deep winter in the Olympics.

Of all the sightings that took place in 1974, probably the most extraordinary, certainly one of the most exciting to the research team, was a double sighting that took place in the Hood River National Forest, about thirty miles west of The Dalles, in July.

Jack Cochran, 43, lives in Parkdale, in the Hood River valley, in Oregon. By profession he is a logging crane driver. He is married, with a family. In his spare time Jack is an artist in the field of wood carving and his models of animals, bears, deer and men, are fine examples of the wood carver's art. In his community at Parkdale, Jack is regarded as a man of standing and integrity. Living not far from Jack, on the road between Mount Hood-a small roadside community-and Parkdale, is Fermin Osborne. Fermin is a logger by trade, age about fifty-five, married, with a family, and originally from Tennessee. At this time he had lived in the Hood River valley for about fifteen years and was regarded in his community and by his employers as a man of reliability and honesty.

On the day of the first sighting, Cochran and Osborne were working at a logging operation on Fir Mountain, a 3000-foot hill in the Hood River National Forest, about twenty miles south of the town of Hood River. They were concluding the work of a logging operation that had cleared an area of about five acres. With them at the time was one other person, a young man named J. C. Rourke, aged about twenty-one and also a resident of the Hood River valley.

The weather was clear and the hillside on which they worked, with an eastern exposure, was beginning to take on the full light of the morning sun. Jack's job at this time was to lift broken logs and debris-known in logging jargon as slash-into piles for subsequent burning and at the time of the first incident he was working with his crane and seated in a cab with glass windows, about fifteen feet above the ground. There was considerable noise from the crane machinery and from the crash of the falling logs as they were dropped into piles. Jack's companions, Rourke and Osborne, were working on foot to his left and a little behind him.

While the huge steel jib of Jack's crane was swinging across his front, his eye suddenly caught sight of a figure, standing at the edge of the trees that fringed the clear-cut area of the operation, at a distance of about 65 yards. Jack Cochran had sharp eyes, for when he was not woodcarving, or working in the forest, he spent his time hiking and hunting. The figure caught his attention immediately because there should not have been a figure there. One of the things that a crane driver has to be most careful of in his work is people-meaning people on foot-and a professional crane man always has his eye peeled for his co-workers. A false move with the ponderous loads that a big logging crane swings through the air and a man can be crushed in an instant.

So the first thing that Jack did when he saw the figure standing at the edge of the trees, was to look back for his two co-workers. He quickly ascertained that they were both where they should be, behind him and well clear of the crane. Who then was the fourth "man" and what was he doing there? To shade his eyes, Jack swung the boom of his crane up and to the right, blocking the sun glare that was lighting the clear-cut.

Then he saw the figure more clearly and when he did he slowed his engine and stepped out of his cab.

The figure was manlike and was standing quite still, seemingly watching the crane and its operation. It was dark in color and, except for the face, the general pattern of color that covered the whole body was uniform. The body itself was massive, with broad, .muscular shoulders and it seemed to stand about six to six-and-a-half feet in height. Its covering appeared to Jack to be thick black hair and its arms hung by its sides. While he watched, the Bigfoot-which Jack immediately guessed it to be-it turned and walked slowly into the woods. When it turned, it put one arm up against the tree beside which it had been standing, presumably to balance itself. Jack then saw it move into the trees and turn to the right and disappear. He did not see it again. He described its movement as smooth, flowing, like the movement of a big man in fine physical condition. The other two men did not see the creature.

Next day, at about the same time in the morning, the working conditions of the three men were repeated. Jack was in his crane cab and Fermin Osborne was working with his young companion, J. C. Rourke; both of them some distance away from Jack and his machine. This time Jack was keeping a sharp watch on the edge of the trees. He knew what he had seen and he wanted to see it again if he could.

Again it was a warm day and about 11.00 AM the increasing heat of the morning sun eventually persuaded Osborne and his young companion to take a break from their hot and dusty work. They put down their tools and walked towards the shade offered by the wall of forest that surrounded the clear-cut. The place that they chose was about fifty yards west of where Jack Cochran had seen the Bigfoot the previous day.

Osborne was in the lead as they entered the edge of the tree line, with Rourke close behind. Suddenly, directly ahead of him and to the left, he saw a huge shape rising out of a dense clump of vine maple and starting to move away.

From where Osborne saw the huge shape, at the edge of the clear-cut, the ground sloped upward and immediately behind the clump of vine maple, in the dense, uncut Douglas fir growth, the ground was covered with a thick carpet of pine needles and stick debris. The creature moved up this slope and Osborne saw it clearly, for there was almost no undergrowth and the dark, tall, manlike figure stood out clearly to his view. His companion, young Rourke, walking behind him, did not have as clear a view as Osborne did; but he also saw the figure.

Osborne, as befits a working logger from Tennessee-and as we learned in subsequent interviews-was basically a simple man, with a simple man's understanding of nature and things natural. His description of what he saw, amply illustrates this fact. He did not pretend to know exactly what it was that he encountered that day; but he did give us a clear and vivid description.

He described the creature as being about six feet in height, covered with thick black hair and with massive shoulders and body. The legs were very thick he said and, under the hair, he thinks, very muscular.

The creature walked upright. At no time did he see its face, for it walked away from him and did not look back. It walked with a smooth but rapid stride and Osborne was surprised at the speed with which this stride began to take it out of his range of vision and away from him. It made no sound and he did not detect any smell from the creature in the air. It did not run, but within a few seconds it had reached the top of the slope behind the edge of the clear-cut and begun to move down the other side.

The reaction of most people who encounter a Bigfoot seems fairly standard. The usual reaction is one of shock, surprise, often followed by near-panic and rapid flight. That this is quite unnecessary, seldom matters to the witness. That the Sasquatch have never harmed anyone, or even threatened anyone, is quickly forgotten in the sudden shock of the sighting and all that people remember, it seems, is what Hollywood and comic books have told us about gorilla-like monsters that tear men to pieces and carry women off to fates worse than death.

Osborne's reaction was quite different from the norm, and he did something which-in our records at least- no one else had ever done. Seeing that the creature was moving out of range of vision, and wanting to have a better look at it-particularly wanting to see its face-and leaving young Rourke standing open-mouthed at the edge of the trees, he ran after it!

He reached the top of the hill as the Bigfoot started to move into the dense scrub on its other side and, perhaps to vent his frustration at not getting a better look at it, or perhaps to try and make it turn, so that he could see its face, he picked up two big rocks and threw them at it. Sadly, it did not stop, or turn and he did not see its face. Within another few seconds it had passed quickly into the dense scrub on the downward side of the slope and out of sight; he did not see it again.

Later, all three men were questioned about the sighting by investigators from the Information Center accompanied by visitors to the Center, among whom were David and Jane Hasinger of Philadelphia, and Nicholas Bielemeier, an associate of the Center and a professional photographer from Hood River. Jack Cochran, in his account of the sighting, gave a clear and precise description of what he saw and of how he saw it. He was fairly sure that what he saw was a Bigfoot; he stated that he could not imagine what else it might have been.

It was certainly not a bear and it was not a man. And it was not a deer or a mountain lion or an elk, the only other large animals that one might find in the Fir Mountain area. Jack was a man with extensive experience in the Cascade mountain forests and he was not a man to make a mistake about something like this, an opinion which confirmed by many who know him in the Hood River valley.

Young J. C. Rourke told me that he was not sure of what he saw. Interviewed by us he said, in all honesty, that he did not see it clearly enough to be able to state, definitely, that it was a Bigfoot. He did see something, of that he is sure, but what it was he does not know.

Fermin Osborne's account of the incident was straightforward and simple. He did not know that the creature he saw was a Bigfoot. He did not know what it was. In fact the actual words that he used to describe it to us, in our interviews with him later, were, "some kind of a goddam monster." But he is sure that he saw his "monster," that it was there, flesh and blood, real, live and not anything that he had ever seen or heard of in the woods to that day.

Our investigation of the incidents was as thorough as time and money would allow. We went to the area with all three of the men and spent many hours examining it. Then we camped in the area and searched it for a week. We saw where the creature had walked; its huge feet had left heavy scuff marks and deep tear-like indentations in the soil and the pine needles on the slope. We saw Fermin Osborne's boot prints where he ran after it and we found the two hollows in the soil where the two rocks that he picked up to throw at the creature had been embedded. Where the creature had walked, we found small stones partially dislodged and also pressed down into the soft soil by the weight of its soft but huge feet. We also searched for hair on the bark of trees which it had passed, in particular two that were about three feet apart between which it had walked. Sadly, we found none.

We found the incident-the dual sightings-to be very credible and all of us who interviewed the three men, including David and Jane Hasinger, were convinced that they were telling the truth and that they did indeed see a member of the species that has come to be known as Bigfoot, on that sunny summer morning, on Fir Mountain, in the Hood River National Forest.

The year 1974 also produced its quota of footprint findings. During the summer months and into the fall, three sets were found. One set was found by a group of young female hikers in the Three Sisters mountains, in the Oregon Cascades. Another set was found high on Mount Jefferson, just north of the Three Sisters, by an engineer from Pascoe, Washington. A third set was seen and examined by forest workers north of Mount Adams in Washington state. All three sets were reported to the Information Center by associates of the Center, and photographs and casts were examined by members of the Center and also by the Bigfoot Board of Examiners.

There are many more examples of encounters with the Bigfoot. But the ones described in this chapter are of particular interest because in each case the people involved were interviewed by researchers from the Center and in most cases the actual place of the encounter was also searched, examined, photographed and documented. Where the examinations produced evidence that thoroughly supported the account of the witness and if the incident was of recent date, extensive searching was carried out in the area of the incident. In some cases, as in these descriptions, the names of the witnesses were released to the public and the press. In others, where the witnesses did not wish to have their names made public, the policy of confidentiality that is maintained by the Information Center was strictly observed.

As to other reports, not a few of the very many sighting and footprint findings stories that have come into the Information Center in The Dalles are undoubtedly improbable. But many of them appear to be genuine accounts of actual encounters, or real finds. Whatever credibility one may allow them, the odd thing is that there is so little variation, among the accounts, in the basic description of the subject matter which is always of bipedal, hair-covered, man-like primates.

Color description has varied, of course, as have height estimates. But there the variation ceases and one cannot help thinking that if all of the stories, all of the accounts, from the early 1800s onwards, were imaginary or simply fabricated, there would be a wider and consequently less credible area of description. No one has ever seen one walking on all fours, for example. Why not? If the account is going to be of something that one has imagined, why not have him, or it, walking on all fours? Surely the story will be just as believable-probably even more believable if the story teller has his imaginary creature on all fours. Again, why have him, or it, so uninterestingly tame and boringly shy? Why not aggressive and dangerous, charging on sight and uttering savage growls and snarls? Surely a huge, hairy, gorilla-like creature would be expected to act like this and surely this is the description that one's listeners would expect and would find much more exciting in a story about an encounter? If one is going to be so bold as to say that one has actually encountered a Bigfoot, surely one is not going to spoil a good story with a dull and unconvincing account of a shy giant that simply turned and walked away and did nothing. This is not what the modern audience wants at all. This is not the stuff of which legends are made.

But, oddly enough, this is all that the audience is liable to get in an account of a Bigfoot incident. In all the known years of Bigfoot history-and in all of the accounts that are not included in this book but which go to shape the historical background of the phenomenon-all of the stories have a sameness about them, a repetition which, were the subject matter not so interesting, would make them dull reading indeed. Looking back at them one cannot help thinking that the authors of these fanciful stories, creating them as they did out of their own imaginations and with all of the license that imaginative storytelling allows, were singularly lacking in creative zeal. Unless, of course, unbelievable as it may seem, the stories are all true.

CHAPTER THREE

From the Bizarre to the Believable

"I've hunted in the Northwest for more than fifty years now, with much of that as a full time professional hunter for U.S. Fish & Wildlife. And I've always been skeptical about the reality of the Bigfoot phenomenon. But you know, just two weeks ago, I had a Black bear walk past my camp and that thing was a monster, all of 800 pounds plus. I've never seen or heard of anything in the Black bear family of that size before. So it makes you think … of what might be out there, that you might have missed." Steve Matthes (1915-1995) to the author, a few days before he passed away at Carlotta, California.

Dennis Jensen and I had visiting Indian friends in Hoopa, California, and a bright fall day in October, 1971, saw us driving up Highway 199 in southern Oregon on our way back to base in The Dalles. We stopped briefly in the little town of O'Brien, not far north of the California border, where I mailed some letters. O'Brien consists of a post office-cum-gas station, a cafe and a few houses back off the highway. I remembered coming in there once, back in 1960, during the first searches, and stopping for a coffee at the cafe. The waitress was talking about Bigfoot, as were so many people in those days. I asked her if she knew anything about it, or had ever seen one or spoken to anyone who had seen one. No, she had never seen one. No, she had never met anyone who had. And as far as she was concerned the whole thing was ridiculous. Why, she added, there was even some crazy Englishman cruising around the country with one of them science fiction dart guns, trying to find one. She had read about him in the papers. His name was Brin, or Bruin, or Brine, or something like that.

We left O'Brien and drove our worthy Scout north on Highway 199 and that evening made camp on the north bank of the Illinois River in southern Oregon, about half a mile back from the highway in a small sandy flat with scrub timber and clumps of tall river grass.

It was a pleasant campsite with clean fresh water in the Illinois and ample driftwood for a cooking fire and, later, a camp fire. The evening was clear and dry and the air had that crisp feeling to it that says winter is coming. A few deer moved in the scrub around the campsite and the only sounds were the occasional rumble of a truck on the highway and the calling of night birds. Sitting on the tailgate of the Scout we drank a few mugs of Carlo Rossi red wine together, cooked and ate our dinner, and afterwards sat at the fire with mugs of coffee and talked about the Bigfoot mystery.

Jenson-when he worked with me-had more field experience and background than anyone else-both in the Bigfoot field and the people associated with it-and his accumulated knowledge of the general subject was unequaled. He had many interesting stories to tell of his experiences when he worked with Roger Patterson after the 1967 film incident (see Chapter VII) and of his own background. Some of them were quite serious and some of them hilarious.

FOOTNOTE. Dennis Jenson was raised in Idaho, where he had worked as a cowhand and cattle manager. Age 35 at this time, (1975), he has been a field researcher and associate of the Bigfoot Research project since 1971. He formerly worked with Roger Patterson, of '67 Bigfoot footage fame.

Even the serious ones, interspersed as they were with Jenson's escapades with the fairer sex, had their amusing sides. Once, visiting a farmer whom he wanted to interview in connection with a sighting, Jenson arrived at the farmhouse to find it empty. It was a hot summer's day and so, instead of waiting in the sun, he walked into the barn, crawled up into the hayloft, and went to sleep. He was awakened by the farmer's bosomy daughter about an hour later, forking hay down for the cattle and, incidentally, nearly sticking Dennis with the pitchfork in the process. He explained to her that he had come to see her father and talk to him about a recent incident near their farm. The young lady listened to his story and then took him up to the house to meet the old man. The old man, a crusty backwoods farmer of the Washington coast range, took one look at Jenson's Levis, with hay sticking out of them at every point and his daughter's hair, also full of hayseed from the forking, and reached for the family shotgun. Jenson fled.

We talked late that night, hunched over the driftwood fire, and climbed into our sleeping bags on the river sand as a thin moon sank across the Siskiyou Forest to the west. But before the fire died Jenson brought another name out of his hateful of experience. It was the name of a man who lived not far from where we were camped and he thought, as I did when I recalled the name myself, that it might be worth visiting him. The name was Flumpf.

Next morning we set out to try and find Mr. Flumpf. We drove into the town of Wonder and then backtracked to Selma. There was a gas station owner in Selma who knew Flumpf, and from him we obtained directions to the man's cabin in the woods west of Selma. A few hours later we found the cabin and sure enough, Mr. Flumpf was at home.

I forget what it was that we had heard about Flumpf. Something to the effect that he had had an experience with a Bigfoot or that he knew something about them that others did not. In any case whatever it was seemed to be worth investigating and so here we were at his cabin, about three miles back in the Siskiyou Forest.

As we entered the wooden garden gate, a small man with bright piercing eyes, who had been sitting on the porch, came down to meet us. It was Flumpf. He asked us what we wanted and when we told him that we were doing an investigation of the Bigfoot mystery-a full-time serious investigation-he stopped and looked at both of us very closely. For a moment no one spoke. Then he leaned forward and in a near whisper said, yes, we had come to the right place indeed, for no one could tell us the things that he could tell us about the Bigfeet. Jenson and I glanced at each other as we followed Flumpf to the verandah of his house, where he bade us to be seated. Mr. Flumpf then told us about his visitors, the Bigfoots he called them, as he knew them and as no one else did.

The Bigfeet were from Venus, he told us, and they came and went quite regularly in Venusian rocket ships. They came to earth to rest, on a sort of R&R basis and they usually came in small family groups. He, Flumpf, was in constant communication with them, not only in the woods surrounding his cabin but also with the Venusian headquarters and the rocket ship base. As owner of the local land and the only person living in the woods for some miles in any direction, Flumpf had been taxed with the job of looking after the visitors, keeping them happy with small food offerings and protecting them from the curiosity of strangers to the area. The present family, he told us, looking across at the densely wooded hillside opposite his cabin, was a case in point. They indicated that they liked certain types of food and not everything that he left out for them. They had met with him half a dozen times since their arrival and they had explained a difficulty that they had in which he might be able to help. This was to find a husband for their unmarried daughter. She was about eighteen and, he said, just about ready to take a husband. Flumpf had indicated to them his willingness to be a candidate in matrimony, should a suitable mate not be available elsewhere. The prospective bride, he told us, was not uncommonly and, according to Flumpf, had indicated to him her pleasure at the proposal. How she had done this Flumpf did not actually say.

The rocket ships landed and took off on the sand flat in the river below his house. This was also where he met the families and where he put out the food offerings. There was another ship due in very soon and Flumpf was at this very moment standing by for a radio signal from the Venusian rocket base with its arrival schedule.

People have asked me, from time to time, what has been the hardest part of my search, the most difficult time, the one that put the most strain on me. I answer without hesitation, having an Anglo-Irish sense of humor, that it is often in trying to keep a straight face in circumstances like these.

We did, on this occasion, manage to get to the gate and into the car before we collapsed in gales of laughter but Jenson, at other times so dependable, behaved quite badly on this occasion. Somehow, while we were listening to Flumpf on the verandah, he slowly managed to move his chair, inch by inch, until he was in a position behind and to the left of the speaker. From there his face could not be seen by Flumpf but it could be seen by me and while Jenson closed one eye, both eyes, scratched his ears, pursed his lips and contorted his mouth in a serious of apelike silent guffaws, I, with Mr. Flumpf looking straight at me, was unable to do anything but keep a serious and supposedly interested face. It was most unfair.

We concluded our visit to Mr. Flumpf with what we thought was a professional touch. We did go down and look at the sand bar behind his house, where the rocket ships allegedly landed. Alas, we found no Venusian footprints, no scorch marks from flaming rocket tubes, and no outer space beer cans or other galactic debris.

Albert Ostman, of British Columbia, has a rather different story than Flumpf to tell of his adventures with the Bigfeet. Ostman insists that he was picked up and carried away by a Bigfoot, to a Bigfoot lair, one night in 1924.

Albert Ostman is now in his mid-eighties. At the time of his experience he was thirty-four. He was a logger, or lumberjack, and indeed this is the trade that he followed all his life. In that year, 1924, he decided to take a little vacation and go prospecting. He chose Toba Inlet, one of the deep inlets of the British Columbia coast and he went there on a Union Steamship boat, disembarking at the little coastal township of

Lund. There he engaged the services of an Indian who paddled him by canoe to the head of the inlet.

The Indian told Ostman many stories about gold brought out by a white man from a lost mine that supposedly existed at the head of Toba Inlet. This white man spent a great deal of time drinking in saloons, but he never seemed to run short of ready cash; he would just make a short trip to his mine and come back with bags of gold. However, he went back to the mine once too often, and he was never seen again. There were those who believed that a Bigfoot had killed him. Ostman had never heard of Bigfoot at that time, so he inquired of his Indian guide what they were like. The Indian told him, "The Sasquatch are big people living in the mountains."

Ostman was not inclined to believe the Indian's story, putting it down to legend and superstition, but the Indian assured him that, though the Bigfeet were few in number, they did indeed exist.

Ostman and his guide made camp at the mouth of a creek at the head of Toba Inlet. The Indian ate with him, but insisted upon setting out for Lund that same night, with the tide. Ostman told him to return in three weeks' time, that he would be waiting in the same spot.

Next morning, Ostman set out on foot in a northeasterly route, looking for a deer trail into the mountains. He took his 30-30 Winchester rifle, a pick that had an axe on its other end, and his pack. The pack contained cans of sugar, salt, matches, a side of bacon, a bag of beans, some prunes and canned goods, flour, hardtack, a quart sealer of butter, three cans of snuff, and shells for the rifle. He cached some of his food so that he would have it when he returned, rolled up his sleeping bag and cooking utensils, and set off. He hiked up through the mountains until mid-afternoon, when he came to a place that was flat, sheltered, and had good water. It was quite high up, and he had a beautiful view out over the foothills and the water of the inlet. He prospected there, but found nothing.

The next day was like the one before. In an area that seemed to offer good prospecting possibilities, Ostman found a suitable campsite and decided to spend several days there. He had just settled down in what he thought might make a good base camp when, as he recounted later, things began to happen.

The first night, his equipment was disturbed, though nothing was taken. Ostman dismissed the disturbance as the work of a porcupine, and thought no more about it. He went prospecting during the day, shot and cooked a squirrel for his supper, and settled down for the second night.

In the morning, his pack had been emptied out, and some prunes and pancake flour were missing. Now Ostman was curious. He found no tracks, but he knew that the culprit could be neither a porcupine nor a bear, since a porcupine could not have opened his pack, and a bear would have made more of a mess.

The third night, Ostman determined to find out just what was, as he put it, playing merry hell with his belongings, so he arranged everything just so, closed the pack, and climbed into his sleeping bag fully dressed, except for his shoes, which he put into the bottom of the bag. He kept his hunting knife with him in the bag and his rifle close by, fully intending to stay awake to catch the intruder.

Some time during the night, Ostman said, he was awakened by something picking him up, literally bodily, and carrying him off, still in his sleeping bag. He was unable to get out of the bag, nor could he reach his knife and defend himself. He could feel the rise and fall of the ground, so that he knew when his captor was going up or downhill. At one time, Ostman was dragged along the ground. He was carried thus for about three hours. Then he was unceremoniously dumped out on the ground and, emerging from the bag, he found himself confronted by four large Bigfeet, which stood around staring at him and chattering to each other.

At first, while it was still dark, he could not see them clearly. Then, as the dawn came up out of the east, he was able to examine the creatures. The Bigfoot group, according to Ostman, consisted of a large male, a large female, and two smaller specimens that Ostman called children. They looked like humans to him, rather than apes. In fact, Ostman used the word "people" in referring to them. He thought that the young ones, a "boy" and a "girl", seemed frightened of him and that the large female did not seem overjoyed to see him.

The male, he says, was about eight feet tall and barrel-chested, with huge arms and legs. He had what seemed to be exceptionally long forearms and large hands with short, thick fingers. The fingernails were short and broad and the creature was covered with hair all over, as were the other members of the family. The only parts of the body that were bare of hair were the soles of the feet, the palms of the hands, the nose, cheeks and eyelids. He estimated the weight of the male at over 700 pounds, and the female at close to 550 pounds.

Ostman stayed with the family for several days. The male had taken Ostman's pack along, so he had plenty of food (including all three cans of snuff) and he was quite comfortable, except for wondering how he was going to escape from his captors. He was able to walk around the Bigfoot habitat-an area of about eight or ten acres, Ostman thought-and he discovered the sleeping quarters of his jailers. It was like a cave in the side of small cliff, the floor of which was covered with moss; Ostman could see several items that seemed to be blankets, woven of cedar bark and packed with dry moss.

Soon, Ostman said, he began to make friends with the young ones, in hopes of enlisting their assistance in an escape attempt. By the sixth day, the young male was quite at ease with him, watching him cook and eat his food, and showing quite a bit of interest in his snuff-taking. The creature weighed about 300 pounds, and had a large chest; Ostman estimated it to be fifty to fifty-five inches. His jaws were wider than his forehead, and his hair was about six inches long all over. He was very agile; one of his favorite pastimes was to grasp his feet with his hands and bounce playfully along on his hind end, apparently attempting to see how far he could go.

At last the older male, who had prevented Ostman from leaving at one point, began to show some interest in Ostman's snuff. One morning, after watching Ostman take it for some time, he reached out and took the snuffbox from him. Apparently thinking that it contained food, he emptied its contents into his mouth and swallowed it.

Within a few minutes, the creature was in extreme discomfort, rolling on the ground and screaming in pain. Ostman determined that this was his chance for escape and, grabbing his rifle, he made for the opening in the canyon wall. The female attempted to come after him, but he fired a shot over her head and she ran back to the lair. He got away and, after several days of wandering, managed to reach a logging camp on what is called the Salmon Arm of Sechelt Inlet. From there, helped by the camp's logging crew, he was able to make his way back to Vancouver.

There was a great deal more detail in Ostman's original account of his experience and, to my way of thinking, the more detail there is, the more believable the story. Many years after the incident, in 1957, he was asked by Canadian authorities to make a statement before a Justice of the Peace, one Lt. Colonel M. Naismith, at Langley, B.C., and he did this, to support his story and to convince others that he was telling the truth.

He was then interviewed by the same magistrate, who, in a separate statement, said that he found Albert Ostman to be in full possession of his mental faculties, of pleasant manner, and with a good sense of humor. Naismith added that, after examining Ostman, he was left with the impression that Ostman certainly believed the story himself and that his examination and cross-examination failed to bring out any evidence to the contrary.

Is Ostman's story true? Personally, I think that if we are to allow that the Sasquatch exist, then it could be. The discovery of a live Bigfoot that matches Ostman's description of his captors will help to prove that he has told a true story. Curiously enough, his description matches very closely the figure that is seen in the 1967 footage made by Roger Patterson. And it was given by him, in writing, in 1957, some ten years before the Bluff Creek filming. Readers will also notice the similarity of his account with that of Muchalat Harry, whose story is the one that opens this book.

CHAPTER FOUR

A Nervous Afternoon for the Welch Brothers, a Disturbed Meeting at the Dalles, a Nasty Fright on a Lonely Road.

When you have eliminated the impossible, whatever remains, however improbable, must be the truth. Sherlock Holmes, to his immortal companion, Dr. Watson.

In the preceding chapter we have looked at two Bigfoot stories. The first one is usually good for a laugh or two when told in detail. The second is different and while its age now precludes a thorough investigation, all that I can say is that people who have interviewed Albert Ostman believe him to be telling the truth. Also, as mentioned above, Ostman, on request, signed an affidavit in the presence of a magistrate-something that in Canada is viewed as very serious-who subsequently noted that he appeared a sane and sensible person.

In this chapter we deal with three more stories. They are simply three accounts of sightings that I have picked out of the many hundred that are on file at the Information Center in The Dalles. All three have been exhaustively checked, and interviews with most of the people involved in them have been recorded. This comprehensive checking system, a part of the daily routine of the Information Center, in these three cases gives them a ring of truth that is hard to ignore.

The first story is about two mining prospectors, Canadians, who in 1965 found themselves, while searching for minerals, in wild mountain country at the head of Pitt Lake, in British Columbia. Ron and Loren Welch worked for a large mining concern and their job was to locate mineral sources and then stake claims on them for their company.

The country to the north of Pitt Lake is rugged and inhospitable. Man stories have come out of the area of lost mines, of impassable gorges, and of men who have gone in there and disappeared, never to be seen again. Pitt Lake itself is a body of water subject to sudden and violent storms and it has drowned many people. Generally speaking it is not a country for amateurs.

But the Welch brothers were not amateurs, they were professional prospectors and their knowledge of survival in the wilderness was born of many years of experience. They were not the kind of men to go into the British Columbian wilderness unprepared, just as they were not the kind of men to imagine something like a Bigfoot. Theirs was a hard-headed approach to a profession that did not allow for flights of fancy, did not allow for very much more than hard work and plenty of it.

One afternoon they were trekking northeast, close to the shores of a small lake. There was snow on the ground and the lake was frozen. There was light timber in the area. The weather was clear, with a thin sun and no wind. They stopped to rest and have a smoke, taking off their packs and placing them against a rock. They sat for a while and were smoking and talking quietly when one of the brothers noticed a movement among the trees about a hundred and fifty yards away. Both brothers looked up and saw, standing and watching them, a very large, hairy, dark-colored man-like figure.

The figure, they said, simply stood and watched them and they gained the impression that it was as curious about them as they were about it. The only movement that it made was with its arms, which hung down by its sides and which swung slightly as its body swayed a little. It seemed to be about eight feet tall and very heavily built. It also appeared to have almost no neck. It was thickly covered with hair, all over, and the color was a uniform dark brown, almost black. The face seemed to be hairless, but it was difficult to be sure at the distance at which they saw it. The backs of the hands seemed to be lighter in color. They were either less hairy or the actual hair on the backs was a paler shade.

The brothers kept quite still and they say that they watched the creature, as it watched them, for probably several minutes. Then suddenly it turned and walked away, its massive body quickly disappearing into the trees. They did not see it again but later that day, when they came back through the same area, they were nervous at the thought of meeting it and instead of taking the same route back, they walked out on the frozen ice and down the middle of the lake. The ice was thick enough to bear their weight but not, they hoped, strong enough to bear the weight of whatever it was they had encountered.

Next day, still working in the same area, they found a set of footprints. The prints were old and melted out to a length of twenty-three inches. They found a second set of ten inches in length. The second set came and went from a small lake where there was a big hole; the ice had been broken and the snow swept back. The hole was four or five feet in width and quite like another I found in northern Washington in 1971.

The Welch brothers returned home and for a while kept quiet about what they had found. They knew that people would view their story with great skepticism. But it was difficult not to talk about it, and after a while their story leaked out. For a few days nothing happened. Then the Vancouver Sun , a large-circulation newspaper of some influence in British Columbia, called one of the brothers about the sighting. Both brothers had agreed not to talk to the media and now a problem arose about publicity. Eventually the persistent Vancouver Sun agreed to a compromise. If they could have the story they would not print the brothers' names.

They got the story but their reporter who wrote it up quite frankly did not believe it. The brothers suggested that he fly out to the area and have a look at the footprints but the newspaper balked at the cost of chartering a helicopter. Annoyed at the skepticism of the newspaper and its staff, one of the brothers offered to pay for the charter of the helicopter. The Sun agreed and a few days later one of the Sun's reporters flew with one of the Welch brothers to the area. There he and the helicopter pilot saw the footprints. By this time they were melted out and deformed by wind and sun effect. But the reporter returned satisfied with his story and satisfied with the integrity of the two men. The Sun subsequently published the story and there the matter came to an end.

To now, the Welch brothers have never returned to the area of the sighting, and to the best of my knowledge, there have been no organized expeditions or searches for what they saw in those rugged ranges of the Garibaldi mountains, to the north of Pitt Lake.

The Dalles is a town of some ten thousand people tucked into a bend of the Columbia River, in northern Oregon, about eighty miles east of the city of Portland. It was originally called Les Dalles. The word "*dalles*" is French and means stepping stones, or flagstones, and it presumably referred to the big flagstone like rocks that rise out of the bed of the Columbia River at this place and which can still be seen today, just below The Dalles dam. The name was given to the area by early French settlers and traders working the Columbia River. History records that the first settler in The Dalles was a Frenchman named Joseph Lavendure. He arrived in 1847 and left in 1848 to settle in California.

The Dalles is the county seat of Wasco County and so it houses all of the bureaucratic offices that are normally found in the principal town of a medium-sized county, including state, county, and city police. There are the normal number of schools, shops, clubs, churches, garages, and motels that one finds in the average American town of this size. There is even that fast-disappearing institution, a shoe shine shop, owned and operated by an elderly Greek who dispenses a mixture of gossip, wisdom, and small philosophy for all comers at twenty-five cents a time.

The Dalles has a reasonably moderate climate. But summers, with hot winds, can climb into the hundreds and stay there for days and winter snaps can be cold and wet. Lewis and Clark, the explorers, and David Thompson, the famed Northwest Fur and Trading Company, passed through the present site of The Dalles in the early 1800s but there were no permanent settlers in the area until 1847. After the establishment of a fort-Fort Lee-a town began to grow and by 1852 there was a small and expanding community. A charter was granted to Fort Dalles, as it was called, in 1857. Soon after the name was changed to Dalles City and later still to The Dalles. The city had the first newspaper between the Missouri River and the Cascades, the Journal and by the late 1800s, The Dalles was a flourishing community. Today the city is pleasant and quiet. The town has many old wooden houses that lend it a certain charm and dignity and the mighty Columbia, flowing quietly through the bend on which the town is situated, reminds one of the colorful past that is the history of northern Oregon. There are a few Indians in The Dalles now, but the tribe-the Celilo-that once lived on the bend of the river and fished its waters for salmon and sturgeon are almost all gone. They called the Columbia the Wauna, a beautiful word that is somehow more fitting than its present name. In years gone by the Celilo fished the Wauna with hand nets and salmon was a major part of their diet. The building of the dams at Bonneville, Cascade Locks and The Dalles put an end to their livelihood on the river.

The city of The Dalles lies on the northern border of Oregon, and the Columbia River is the borderline between Oregon and Washington. The area of Washington that lies to the immediate north of The Dalles is mostly open pastureland with sweeping hills that climb west and north to the Cascades. Almost due north is Mount Adams, a mountain that the Indians called Klickitat. The name means "galloping horse" and said quickly, three times, the word does have the ring of galloping hooves. The ranges that run out of the southern watershed of Klickitat are called Rattlesnake Ridge, Toppenish Ridge, and Horse Heaven Hills.

East of The Dalles is generally open country that runs all the way to the Wallowa Mountains, the onetime home of Chief Joseph of nineteenth century fame. To the south is more open country, wheat growing and farming land with a scattering of small towns that include Dufur, Shaniko, Antelope, and Madras. To the west runs the gorge of the Columbia and it is a deep and spectacular gorge, with three to four thousand foot walls in places and many beautiful waterfalls, the long white plumes of which tumble through the dense green foliage of the hillsides. The gorge is also a funnel for the wind that pummels The Dalles, summer and winter, to where some old timers jokingly it "Windy City II."

At the western end of the city is a long, flat hill, about 800 feet in height, variously known as Strawberry Hill, Raspberry Hill, or Table Mountain. Half a mile short of this hill the city comes to an end in a sprawl of trailer courts and sales lots, garages and some farmland. The eastern slope of this hill contains a quarry owned by Arlie and Jack Bryant, father and son partners in a rock crushing business. On the west side of the hill is an abandoned asphalt plant owned by two brothers, Howard and Stanley Stinson. Land on the top of the hill is divided in ownership between these two families and a third owner, Ernie Kuck. The land to the west, below the hill, land that stretches for a thousand acres west and south, is owned by a Mr. and Mrs. William Marsh. The Marshes, long-time settlers in The Dalles, with an ownership going back several generations, call their land Hidden Valley and their one thousand acre parcel is composed of basalt rock and stunted oak tree country, with short grass and shallow ravines that hold deer, pheasant, quail and, occasionally, bear. There is some water, here and there, in small springs.

Fringing and enclosing the northwestern edge of this whole area is a hill that is called Crate's Point. It is named after John Crate, one of the first settlers in the area, a Hudson Bay Company man who married an Indian girl and by her fathered a large family, some of the descendants of which still live in the area today. This hill begins at the river, crosses The Dalles-to-Portland freeway, crosses the old highway, and then climbs rapidly to a height of 2020 feet. The lower reaches of this hill are rocky and bare, with sparse grass and scattered small oaks. There is much rock, broken and tumbled, and in places there are narrow, shallow ravines. On the western edge of the hill there is a small cliff that drops sheer to the old highway, where, on a series of man-made terraces, lie a dozen or so trailer homes. A few pines grow in this area and their twisted and bent shapes are an indication of the force of the wind that assails the whole area for much of the year.

Older maps of the area show a small town, high up on Crate's Point Hill, called Ortley. But Ortley is long gone, deserted, and today there is nothing left of it but a couple of buildings used for machinery storage by a Mr. and Mrs. George Johnson, who own the hilltop land. Old-timers say it was the endless wind, cold in winter and hot in summer that drove the people of Ortley to abandon their homes there in favor of a more comfortable climate.

Between the land enclosed by Crate's Point Hill and Table Mountain and roughly facing the center of it, situated on the old highway and lying between that highway and the main freeway, is a trailer court. It is called the Pinewood Trailer Court, and it consists of a couple of dozen old trailers, a swimming pool, some ragged flower beds and an office.

In midsummer, 1971, on a fine afternoon with bright sunshine and a clear sky, three Portland businessmen-owners of the trailer court-were having a meeting in the trailer court office. The time was about three-thirty and all three were deeply engrossed in the affairs of the trailer court, when one of them happened to look up and see what looked like a very big man, covered with hair, walking among the rocks in the meadow directly opposite the office. It was, apparently, clearly visible through the trailer window.

The meadow that lies opposite the trailer court is about 400 yards in length, roughly east to west, and 200 yards in width, roughly north to south. The old highway, also called the Rowena Highway, borders the north side of the meadow and the south side is bounded by a small cliff of broken basalt rock about fifty feet high. The area has thick grass and a scattering of small oak trees. The rock cliff is broken in several places that could allow access for the nimble-footed to the upper meadow, which stretches all the way south into Hidden Valley. The cliff itself comes to an end almost opposite the trailer court office and it was at the end of this cliff, walking among the broken rock, moving slowly, seemingly slightly bent over, hesitating, appearing, disappearing, and finally going out of sight behind the end of the little cliff, that the three businessmen from Portland saw the figure in question.

The Sheriff's Department of The Dalles, headed by a man of many years' law enforcement experience, Sheriff Ernie Mosier, is an efficient, well-run organization. In the matter of the Bigfoot phenomenon and my investigation, headquartered in The Dalles, I would describe the general attitude of the department as interested, helpful, cooperative; and open-minded. Among the deputies who took a particular interest in the Bigfoot phenomena and in our investigation in 1971 were Richard Carlson, Jack Robertson, Larry Tillinghusen, and Bob Hazlett. The man usually assigned by the department to investigate reports of Bigfoot footprints and sighting claims was, in midsummer, 1971, Deputy Richard Carlson.

Carlson was intensely interested in the Bigfoot phenomenon and this interest spurred him to offer and supply much help to my investigating team. Backing the interest was an open-minded attitude toward the phenomenon, an attitude that befitted an intelligent officer of a well-run law enforcement agency.

On June 2, 1971, Carlson wrote a routine report for the Sheriff's Department. The report describes how Carlson, after hearing of the sighting supposedly made by the three men, called them in Portland and by phone spoke with Dick Ball, Jim Forkan, and Frank Verlander. All three made statements to the effect that they had seen what they thought was a Bigfoot. Verlander in particular was adamant about what he saw. He told the deputy that he definitely saw it, that it was very large, very wide, and probably close to seven-and-a-half or eight feet in height. Forkan and Ball confirmed the story.

In this same week, in The Dalles and more or less in the immediate area of Crate's Point, other people made similar claims of sightings. There was a maintenance man at the trailer court, one Joe Mederios, who said that he saw it twice.

There was a music teacher and his wife, Mr. and Mrs. Richard Brown, and two others. All gave very much the same description, the only difference in agreement being on height. But all said that it was very tall, and no one thought that it was under six-and-a-half feet.

One of the part-sponsors of my investigation at this time was a man Tom Page of Mentor, Ohio, a businessman and member of the Explorers Club. Page was intrigued by The Dalles sightings to the point where he decided to fly out and take part in the interviewing of the people concerned. He called to tell me that he was bringing his attorney with him, a Mr. Wayne Newton, a criminal trial lawyer, also from Mentor, Ohio. Page wanted to have Newton talk with some of the witnesses and gauge the credibility of their stories.

The three of us, with Dennis Jensen, drove to Portland to talk with the owners of the Pinewood Trailer Court. Frank Verlander, for reasons of his own, refused to see us, but Jim Forkan and Dick Ball allowed us several hours of interviews and were most cooperative and helpful in telling us their stories. We interviewed each man separately and taped most of their conversation. Wayne Newton did most of the questioning and proved himself to be an astute, quick-thinking, and highly intelligent man. While there we also interviewed the Browns, who at that time lived at the Pinewood Trailer Court. And before Page and Newton returned to Ohio, we managed to interview Bob Gimlin, partner in the 1967 footage event.

Later we sat down to analyze the results of our talks with The Dalles eyewitness group. We studied the notes we had taken during the interviews and discussed the sightings for many hours. The result of our discussions was that we, all four of us, were reasonably convinced that Dick Ball and Jim Forkan and their companion Verlander-although we did not personally hear his account-saw something that might well have been a Bigfoot.

The "something" that they saw was man-like in appearance, walking upright, seemingly quite large, grey or dark grey in color. They saw it at a distance of between 260 and 270 yards and they watched it for approximately twenty seconds. The sighting was in good light; both Forkan and Ball had some experience in hunting and with Pacific Northwest wildlife. When interviewed they were quite sure, for instance, that it was not a bear that they saw, or any other animal. Again, their experience of the outdoors persuaded them to admit, in all honesty, that they were not absolutely certain what it was that they saw. It looked to them like a Bigfoot or at the least what they imagined a Bigfoot would look like. But we were quite satisfied with the truthfulness of their story, even if the non-availability of one member of the party, Verlander, did detract somewhat from its completeness.

Tom Page and Wayne Newton were not able to interview Joe Mederios. (He had left the trailer court when they arrived in The Dalles.) But I had seen him and interviewed him and was satisfied with his truthfulness; I believe that he too saw something in the meadow opposite the trailer court, few days after the first incident.

They were, however, able to interview the Browns, with whose story Newton was not completely satisfied. There was some detail in it that he did not like, detail that did not fit in with the sighting; to him it lacked credibility. But Page and I were reasonably well satisfied, as was Jenson. We believed their story to be truthful and we think that they gave us a true account of having seen something. Whether it was actually a Bigfoot that they saw, is another matter. One thing that Newton did not like was that their sighting took place in the late evening, with poor lighting conditions that could have induced error.

The first year of my investigation, my team was based at Evans, Washington, close to the Canadian border. For the second year we moved to The Dalles, in Oregon, not so much because of the sightings that supposedly had taken place in the area as for its strategic location. The Dalles is on the edge of the main Cascade range, and through a network of roads the range is accessible to the north, west, and south. The town is also on the extreme northern border of Oregon and is thus equidistant between the state of Oregon, to the south, and the state of Washington, to the north. It is also, because of its position, halfway between British Columbia to the north and northern California to the south, all supposedly Bigfoot habitat, as suggested by evidence, i.e. sightings and footprint finds.

Geographically, The Dalles made an excellent base for an investigation designed to cover all of this vast area. As to the sightings that were claimed in its immediate area, what we were concerned with and what we considered a most important part of our investigation was the reason, or reasons, why a creature like a Bigfoot would come there in the first place. All wild creatures have a natural reason for whatever they do and the reason is usually a simple one, like the need for food, seasonal migration, mating urges, et cetera. We felt that if indeed it was a Bigfoot that the various witnesses claim to have seen in area of The Dalles, that there must be a very good reason for its coming there. We also felt that if we could discover this reason, that we would be making a great step forward in our search for the truth.

There is some food in the general area of Crate's Point (the area of the sighting claims) in the meadows there and higher up on the hill itself, Jenson and I found wild cucumbers growing in patches in June and

July of 1971. There were also wild potatoes and wild onions on the hill. But there was very little else and what there was hardly seemed enough to attract a large primate.

We looked for other reasons. An ancient river crossing place? An ancient route, from the south, along the ridges that run east of Mount Hood? A migration route? A meeting place? A mating area? None of these seemed to fit and the reason, was, simply, the character of the vegetation of the area.

All wild creatures need cover. They need it as much as they need food and water. The mouse has his tiny tunnels and grass-covered runways. The elephant has his forests. The tiger his grass jungles. In particular, large primates need cover and one has only to look at the habitat of gorilla, or chimpanzee, or orang, to see this. But at The Dalles, at least the area where these sightings supposedly took place, there is almost no cover. The forests that roll all the way from the coast, the great stretches of coniferous trees, come to an end at the top of Crate's Point Hill. From there on, down across the meadows of Hidden Valley, up over Table Mountain, into The Dalles itself and then on beyond The Dalles for more than a hundred miles, the land is almost bare of cover of any kind. There are ravines with scrub timber, patches of oak and small trees. But there is no real cover, the kind that a large, shy, wild creature needs.

As to the possibility of routes, the Columbia River, in The Dalles area, does have something to offer and this is that in this area the river narrows and thus offers easier crossing for something that is going to have to swim to get across it.

There are many other factors that have to be taken into consideration and all of them have a place in the "Why The Dalles?" Puzzle, a- puzzle that still, after all this time, has us wrinkling our brows. My friend Jensen's jocular solution, "It's probably just sentiment. The things like to come every year and sit and look at the river from Crate's Point," I find somewhat unsatisfactory.

The third story takes us to British Columbia, and thinking of the location involved, I am taken back to the high ranges of the Nepal Himalaya. Deep in the heart of the peaks that the Nepalese called the Mahalangur Himal-the mountains of the great monkeys-there is a high and permanently frozen pass. The wind whistles through it for three hundred and sixty-five days of every year and the temperature never rises far above zero. A British expedition, with grim humor, named it Brass Monkey Pass and if the reader does not know why they gave it this name, I suggest that he take a trip there and spend an hour or so on the pass. He will quickly find out.

The area we are looking at here however, in this story, is not the Himalaya but the Cheakamus River in southwestern British Columbia where, if its winter and you have been huddled for ten hours on some bleak rock, watching for a sign of a Bigfoot, you will see the analogy with what the British were thinking when they named their Himalayan pass.

As any outdoorsman knows, each mile per hour of wind speed means one degree less of temperature. When the temperature is already close to zero and the wind is blowing twenty knots, only bears, madmen, and Bigfoot hunters go out in the midday freeze. The weather was like this when I first went to the Cheakamus, in December 1971, to examine the place where William Taylor, highway maintenance foreman for the Squamish area, British Columbia, almost ran into a Bigfoot one bitterly cold afternoon in the early winter of that year.

Taylor has now been transferred to another area of British Columbia, but at the time of the incident his headquarters and his home were at Pemberton, about a hundred miles north of Vancouver. On the afternoon in question, at about three o'clock, Taylor was driving north towards Pemberton. He had been down to meet a man coming up from Squamish with spares, and was now on his way back to Pemberton. About six miles south of Alta Lakes, the main Pemberton -Vancouver highway runs east of the Cheakamus River and about a thousand feet above it. Along the river, also on the east side of it but much lower and closer to the river, runs a railway line. There is a sharp curve in the road, below which the railway line crosses the Cheakamus and about half a mile south of this curve there is a man-made cleft in the hillside through which the road runs. Taylor had driven through this cleft and was coming out of the curve when he saw the Bigfoot.

His first impression was that it was a bear, for it was down on all fours close to the bank that edged the road on the left, or west side of it. Then it stood upright and as he drew his truck to a sudden halt, it walked quickly across the road to the opposite bank. It climbed this bank, watching him as it did so. Then it went over the top of the bank, which is about twenty feet high at this place, and, walking upright, disappeared.

Before he was transferred from the Squamish area I went to see Taylor at his home in Pemberton. I found him to be an intelligent, rational man of obvious credibility. He told me his story in direct, simple fashion. He told me exactly what happened and nothing more, and I found it very difficult not to believe him. He wanted no remuneration for the story and no publicity locally. His home, a large trailer unit, was clean and tidy and his family impressed me with their pleasant and friendly attitude. Later, without Taylor's knowledge, I checked on his background with local Canadian police, his employers, and some of his friends. All said the same thing. Good working record, good family man, sane, sensible, no abnormalities and not given to flights of fancy.

Taylor told me that the creature that he saw was at least seven feet tall and probably weighed 300 to 400 pounds. It was of massive build and had a large, protruding stomach. It was dark reddish brown in color and completely covered by hair with the exception of the face. He guessed that the hair on its body was four inches in length. It was about a hundred yards away when he first saw it and by the time he brought the truck to a halt, it was about thirty yards from him The time was early afternoon, about two-thirty, and the day was clear, with light overcast and no sun. It was very cold and he thinks that it was well below zero in the exposed area of the road above the Cheakamus River.

He believed, he said, that the creature must have fallen off the bank, or maybe jumped down and landed on all fours because it seemed to rise awkwardly when it got to its feet. It walked upright all of the time that he watched it and only once, when it was climbing the steep bank on the east side of the road, did he see it put a hand down. It seemed to do this to steady itself, as a man may do when climbing a steep slope.

It did not run, but walked rapidly across the road and it kept its head turned towards Taylor and its eyes on him until it reached the bank. Taylor described the look as menacing (as did Patterson of the Bluff Creek Bigfoot he said he saw) and although he was perfectly safe, in the steel cab of a big solid truck, with both doors securely locked, he said that he felt the hair on the back of his neck begin to prickle and a cold feeling creep over him that was, simply, fear. He does not attempt to hide this. He quite frankly admits that the huge creature that he saw and that seemed to glare at him, as it crossed in front of him, scared him very badly indeed. He wasted no time in getting away from the place and back to his home.

Later his work took him through this area many times, but it was several weeks before he could force himself to get out and walk around on that lonely mountain road where he had seen the Bigfoot.

One aspect of Taylor's story is of particular interest because it gives us a pointer, a clue, concerning the diet of the creatures. The one that Taylor encountered had, he stated, a fish in its left hand. The fish was large-probably a ten pound fish-and the Bigfoot was carrying it with its fingers curled around the fish's body. Subsequent investigation by us of the area, which included climbing down to the Cheakamus River, revealed that a salmon spawning run was in progress there at the time of Taylor's encounter, a run that would have allowed for dying, post-mating salmon to be caught by hand.

Three sighting stories. The Welch brothers at Pitt Lake. The Portland businessmen at The Dalles. Bill Taylor on his mountain road above the Cheakamus River. Three stories of sightings of huge, hairy primates, walking upright, manlike, gigantic, creatures from another world. What are we to believe of them? What conclusions can we draw? What can we really say when we sit down and analyze the stories in all their detail and, more important, the character and the credibility of the storytellers? I think that the reader will agree that there are only three conclusions that make sense. These are, (A) that the storytellers are fabricating their accounts, or, (B), that they were, at the time of the supposed incident hallucinating or (C), that they are telling the truth.

Were they fabricating their stories? It is possible, though not probable, for a man who does this usually has a motive. That motive is sometimes money, sometimes publicity, sometimes private personal reasons that are the creation of warped and unstable minds. None of the storytellers involved in these three incidents seemed to have the faintest motive for fabrication. None of them showed any desire for monetary return for their stories. All of them, without exception, shunned publicity. (The Welch brothers would not even allow their names to be connected with the story when it first appeared in Vancouver newspapers.) And all of them, again without exception, seemed to me, in my private interviews, and to others with whom they talked, to be of sound mental attitude.

Then, were they hallucinating? Did they simply imagine that they saw a large grey, or brown, or black hairy creature, walking around in the middle of the day? Was it perhaps auto-suggestion created by the general publicity that the newspapers had been giving to the public on the subject of the Bigfoot over the years, an inner semi-secret desire to see one, a desire strong enough eventually to produce the mind state that convinced the owner that he was actually seeing a Bigfoot? Somehow, I doubt this. The credibility rating that I would give to all of the people concerned in these three sightings is high. None of them struck me as being men of frivolous mental attitude-of whom I have met a few in my time. All of them, in fact, seemed very much the opposite. In the case of the Welch brothers, their way of life was hardly one that would allow for mental frivolity. Rugged environments have a nasty habit of chastising men who lack mental discipline. As to Taylor, his position of roads foreman and his long record of good service hardly seem to fit with a man who imagined seeing large hairy primates on the highway, unless they were really there. As to Forkan and Ball, an inquiry into their background showed them to be longshoremen, union men, hard workers who made their living in an atmosphere that would hardly tolerate men given to vivid imagination. Their background included the rough and tumble life of the Portland dockyards, a way of life that allows for the survival of only the determined worker and the basic thinker. I think that in their case, as in the case of the Welch brothers and Taylor, hallucination is ruled out. Unfortunately for the skeptics, this leaves us with only one alternative, which is that all of them were telling the truth. Again, unfortunately for the skeptics, I find myself having to accept this conclusion, with just one small stipulation, which is that all of them were hoaxed by some other person or persons unknown. A stipulation that suggests, I must admit, the highly improbable.

Is it possible that they were hoaxed by some other person? Hoaxed by some practical joker who not only kept different colored sets of fur suits in his closet but was also prepared to travel hundreds of miles at the drop of a hat, just for the doubtful privilege and self-gratification of showing himself for a few seconds to some hard-working and rather disinterested eyewitness? Someone who would boat his way up the turbulent waters of Pitt Lake and then, with ingenious navigation, manage to locate himself in the precise place where the Welch brothers were going to be at a given time? A place that he probably did not for sure know himself and which certainly no one else would be able to tell him? (Professional prospectors working for big mining concerns are under rules of the strictest security concerning mining potential claims.) Could it be someone so mentally backward as to find reward in sitting on the edge of a freezing mountain road in the wild hills of British Columbia waiting for a roads maintenance man to come along and have a quick look at him? Could it be someone with such disregard for his own safety that he would walk out in the open fields of The Dalles, in broad daylight, in a country that literally bristles with guns and people who know how to use them-not a few of whom, have told this author that they would use them on a Bigfoot if they ever saw one? In the case of The Dalles open fields that would offer, for a hoaxer, no safe line of retreat and where the nearest cover was over a mile and a half away? Somehow, I doubt it. Thus we are left with what will be, for the skeptics, only one conclusion, which is that all of the eyewitnesses in these stories were telling the truth and that what they saw was a real living creature, the giant primate we call Bigfoot.

I have told these three stories in some detail because they are among the better accounts that the my team of that time produced in its investigation of the phenomena.

There are many other stories and many of them are equally credible. Among them, as interesting examples, are:

A Mrs. Gruber, of Evans, Washington, said she saw one walking on the edge of the highway, near her home, one afternoon in 1968. It walked through the trees, she said, and as she climbed into her car and raced away in fright, it disappeared toward the river.

George Hildebrand, of Republic, Washington, claims to have seen one on the Republic Road, just west of Sherman Pass, on a winter's morning in 1971. It jumped off the road into the snow when it saw his car approaching, a huge, dark-haired creature that looked like a giant man. We got to the scene of this sighting three days later and there were still blurred footprints in the snow where the Bigfoot had plunged off the road. Very deep snow prevented us from following the tracks.

Louis Awhile, of White Salmon, Washington, said he saw one cross the highway just east of Beacon Rock, in Washington, in 1973. His young daughter, who was with him at the time, also saw the creature and was badly scared by it.

Mr. and Mrs. Martin Heinrich, of Portland, described seeing one on the Lewis River in 1968. They were fishing from a boat at the time and watched it for several minutes before it saw them and walked quickly into the trees.

Frank Luxton, another British Columbia highway maintenance, foreman like Bill Taylor, said he saw one cross a lonely road north of New Hazelton, in British Columbia.

There are many other sighting claims in our files of those days and while some of them were not sufficiently documented and others are obviously imagination, or honest mistake (bears, tree stumps at night and so forth), too many of them have the ring of truth to be ignored.

Skamania County, in southern Washington State, has officials who are the first to take a long and serious look at the incidents that were claimed for their county as well as the credibility and responsibility of the people involved in them. The result was a county ordinance that imposed a fine of up to $10,000, or imprisonment, or both, for anyone attempting to injure or kill a Bigfoot, a creature that they believe should have been listed as an endangered species.

CHAPTER FIVE
No Place to Hide

Life is short. Grab it while you can. Work hard. Play hard. Don't complain. Don't explain. Remember that once you're gone, you're gone. There's no coming back. TOM SLICK, to the author, over a couple of glasses of scotch at a campfire at Louse Camp, in the Bluff Creek watershed, fall, 1962, a few weeks before he

I am constantly amazed at the number of people who approach me and tell me, quite seriously, that there is absolutely no such thing as a Bigfoot, and the simple reason for this-surely you should have learned this by now, Mr. Byrne?- is that there just isn't anywhere for them to hide. The Pacific Northwest rivers are alive with fishermen. The hills are full of deer hunters. The mountains are full of elk hunters, bear hunters, cougar hunters, and foresters and loggers and timber cruisers and ecologists and Boy Scouts and Girl Scouts and trekkers and campers and cross country skiers and rock climbers and horse riders and of course, Bigfoot hunters. All of the available land on which houses can be built has been built on. All of the available land for farming has been farmed. Most of the timber has been cut down and between the logging companies and the U.S. Forest Service there just isn't anywhere, anymore, where there is not a road with a truck or a pickup roaring up and down it. The high peaks all have fire lookouts, for fire control. The high ridges all have telephonic relay stations. There is a radar or radio or electronic device of some kind on every hilltop. Planes of all sizes fly across the mountains all the time and those that are being used on fire control work contain men who are actively watching the ground beneath them. Some may be carrying wildlife agents engaged in the work of animal counting in which case they will be studying the ground and its wildlife population with care.

Some may be carrying timber cruisers, who will be watching the timber with equal care, evaluating it for cutting. Even private planes, simply passing over the mountains from A to B, often fly only a few thousand feet above the forest, at altitudes that allow clear views of the treetops and, in the open spaces, of the wildlife therein. In other words, and surely you can see this, Mr. Byrne, creatures of that size simply have nowhere to go, nowhere to hide. This proves, as any simpleton can see, Mr. Byrne, that the things simply cannot exist.

Normally the people who expound the nowhere to hide theory do not have the time to hear, or simply do not want to hear, any argument. They have delivered their ultimatum. They have expounded what anyone but a complete fool can see is the truth. That their argument is based on almost total ignorance of the real situation would never, ever occur to them. That they have no understanding of the incredible vastness of the Pacific Northwest, is something that they prefer to close their tiny minds to, now and in the future. That there just might be a Bigfoot and that in actual fact the creature might have a very large habitat perfectly suited to its needs, is something that they would consider quite unfair to have proved to them.

How is it, then, that wild animals can occupy an area and remain so well hidden for so long? What factors are involved that enable them to do this? There are many, but the principal reason is that they simply do not want to be found-they do not wish to have contact with man. With most wild animals, of course, there is motive behind this and that is that they are hunted by man, as a result of which man is regarded by all of wild animals, certainly all of those who have had contact with him, as a hazardous predator. In fact, no other creature on the face of the earth, since time began, has been so successful in hunting–and eliminating-his fellow creatures. (And his fellow men.) And so all wild animals know, instinctively, that if they are going to survive they must avoid man and, to do this, live in habitat that provides them with the cover needed for protective concealment.

There is a saying that you may find what you are searching for if you search hard enough, but that you will not find that for which you are not searching. This saying can be applied to the Bigfoot to where one might say, in regard to the reason why one has never been found, is that, with the exception of my own research projects, no one has ever searched for the creatures on a full-time basis professional basis.

The finding of the Tasaday tribe, the Philippines, in 1972, is a case in point. This incredible stone-age group was not discovered simply because no one was looking for them. They in turn were not really hiding.; if they were, they would have taken steps to ensure that they remained hidden.

Another example, though on a somewhat different scale, is the amusing story of the Portland Park Hermit. In 1969 two boys noticed a man going into a clump of bushes in Washington Park, Portland City's big municipal park. They followed the man and when he saw them coming after him, he turned and chased them off. The boys informed the park authorities, who soon came to the scene to investigate. In a clump of bushes, in a small island of shrubbery, in the center of the park, officials were stunned to find a cabin, complete with bed and all home comforts. Washington Park has a total area of 149 acres and employs twelve gardeners and maintenance men who work five days a week, nine to five, in the park. There are also several office staff workers and a superintendent. Throughout the year, the park averages fifty visitors a day, though on summer weekends this may rise to as many as three hundred persons. A large number of children also use the park and play in the open grass areas and in the clumps of shrubbery that separate the flower beds and lawns. So the park authorities were not only startled to find a hermit - an elderly man - living in their park in a full sized cabin, but that he had been residing there, quite happily, for fifteen years.

In the southwest jungles of the Kingdom of Nepal, an area where I hunted for many years, there lived a large male black panther. The big cat had as his habitat an area of about sixty square miles, and in this he lived and hunted. In one area, villagers claimed to have seen him several times, both in the dry months of winter and during the rains of the monsoon. They told me about him and I made a mental note that one day I would go after him with a camera. I felt that I would eventually have a chance to see him myself, for I hunted his area for nine months of each year and spent many hours just sitting and watching wildlife, a favorite occupation of mine in those days. I saw his pug marks many times and also his kills, or rather the remainder of them. I found his dens and on numerous occasions I heard him call in the night, the heavy sawing cough that is the hunting call of the leopard.

Had I made a determined effort at searching for him, with either camera or gun, we might have come to grips within a week or so. Had I informed the Nepalese authorities of his presence, they would have hunted for him, for he would have been a prized trophy with his dark green eyes and black velvet coat. But I told no one and for many years he and I shared the same jungles, I publicly and he very privately, very secretly, unseen by the eyes of man. The Kala Cheetua, the natives called him, the black leopard.

Six years after I had first heard of him, I saw him one day, in broad daylight, running for his life from a fast moving grass fire. I saw him as he crossed a clearing at a distance of about thirty yards, bounding through the air in a beautiful blur of motion. He was in sight for about three seconds and then he was gone. I never saw him again, though I lived in the area for many years.

In Idaho, in late 1970, two men were lost in the mountains. There was deep snow and extreme cold and when their pickup bogged down and left them stranded, they found themselves in serious trouble. For some time they stayed with the vehicle. Then one man set off on foot to try and get help. The man who stayed behind eventually succumbed to the cold. The man who set out to try and find help came across an empty cabin and found some food in it and was able to survive. Both men were the subject of a massive hunt in which several hundred men, supported by air search, took part. Every effort was made to find the missing men, and the men themselves, particularly the man who holed up in the cabin, made every effort to contact the searchers and to let them know their whereabouts.

The searchers had a fair idea of where the men had disappeared and so concentrated their search on a comparatively small area. After two weeks of effort that included hundreds of man-hours and miles of footwork, the search was abandoned.

Fifty-eight days later, the man who found the cabin made his way out to a road and was picked up and rescued. Lesson? The northwest wilderness hid both these men and the combined efforts of both the lost men to be found, and the searchers to find them, were in vain.

Since I came to the Pacific Northwest, no less than six planes have been lost in its rugged mountains. In this same period, four planes that crashed in previous years have been found. The high forests of the Marble Mountain Wilderness area, in northern California, provided one find, a small two-seater that had been missing since 1969. Lost in a storm while on a flight from Eugene, Oregon, to Bodega Springs, California, the plane was found one hundred   miles south of where searchers had operated-and given up-some twenty months previously. All hope of finding the craft and its two passengers had been abandoned when hikers in the Abbot Lake area, deep in the wilderness, found pieces of wreckage that led to its discovery.

On December 10th, 1944, a pilot named George Wilbur Carroll took off from Billings, Montana, to fly to Seattle. Carroll planned to be home for Christmas and had loaded the back of the plane with Christmas presents for his wife and children. The machine, a Ryan B-1 Brougham, a sister aircraft to Charles Lindbergh's plane, the Spirit of St. Louis, was last seen flying over Superior, Montana. After that it completely disappeared. For many years it was believed that Carroll had missed the Washington coast in fog and flown into the Pacific. Then, in November, 1971, two U.S. Forest Service workers found aircraft wreckage in a snow-covered ravine near Lookout Pass, on the Idaho-Montana border. In the wreckage they located a number, NC-

6955, and reported it to the Federal Aviation Administration. The FAA identified the plane as Carroll's Ryan Brougham and thus solved a mystery that the wild mountains of the Northwest had kept unsolved for twenty-seven years.

And so, the rivers are full of fishermen. The hills are full of deer hunters. The mountains are full of elk hunters and bear hunters and cougar hunters and so many others that it just is not possible for anything to hide out there or for anything to remain hidden for more than, say, a day or two at the most. You can't tell us, Mr. Byrne, that something like a giant primate could actually find a place to live and hide, in the United States, in this modern day and age?

Invariably, when the subject of the Bigfoot comes up among skeptics, I am also served, in addition to the "no place to hide, nothing to eat, too cold in the winter" etc., theories, two other little gems of modern skepticism. One is the Loch Ness phenomenon. The other is the Yeti, or abominable snowmen of the Himalaya.

On the subject of the Loch Ness monsters I do not have a great deal to say. I have been to the brooding Scottish lake that is the Ness and have peered through the mists like many another, hoping for a glimpse of the elusive monsters. But I have had no personal part in the search and investigations that have been carried out at the lake over the past few years. However, two of the principal searchers are not only friends of mine but also associates in the Bigfoot project and from them I have first-hand knowledge of the work that has been done and which now, in 1975, is continuing at the lake.

The first of these is Robert Rines, President of the Boston-based Academy of Applied Science and head of the American team that has done such sterling work at Loch Ness since 1970. Working with Dr. Rines is an Englishman, Tim Dinsdale, head of the photographic section of the British Loch Ness Investigation team.

I first met Tim Dinsdale at the home of British scientist and monster hunter Ivan Sanderson in New Jersey, in 1972, in the company of old Nepal hands Gerald Russell and Ron Rosner. At intervals in a fascinating set of conversations on such diverse subjects as Yeti, Bigfoot, the Bermuda Triangle, UFOs, Shookpas, and Okopogos, Tim told me of his thirteen-year search for the monsters at Loch Ness, a search that started with a sighting and a piece of now famous 8mm film-taken by him, years previously- and has continued, without ceasing, to the present day. He also told me about the work of the Academy of Applied Science under Dr. Rines and of their findings to date.

In 1970 a team of scientists from the Academy of Applied Science in Boston went to Scotland to commence research at Loch Ness. There they were joined by Tim Dinsdale and together they put into effect a meticulously planned investigation of the thousand-year-old phenomenon, an investigation based on sound scientific principles and professional planning. For the first tests at the lake the Academy used a standard Klein Associates Model Mk-300 Side Scan Sonar. The sonar was used in two modes. One was the conventional towed configuration to examine the bottom slopes of the lake. The other was in the form of a fixed mode, to make positive determination of moving targets.

These first tests produced three important discoveries. They were, 1) that there were large moving objects in the lake. 2) That there was abundant fish life in the lake that could support a large fish-eating creature. 3) That there were large ridges or caves in the steep walls of the lake which could conceivably harbor large creatures.

These tests set the stage for further search and investigation. In 1972 Dr. Rines returned again to Loch Ness. Working with other Academy members he put into action a plan that in August of that year produced their first success, an extraordinary photograph of a section of one of the huge lake creatures. The photograph was supported by sonar findings that indicated that not one but two of the leviathans had swum within range of their search gear. The photograph, which showed a very large fin-like appendage, was later examined by sonar experts at the Massachusetts Institute of Technology and by Woods Hole Oceanographic Institute. The examination showed there were two of the giant creatures and that they were from twenty to thirty feet in length, with several humps and fins and with long tails. This expert examination of the evidence plus the impeccable credentials of the Academy group removed any doubt as to the authenticity of the findings and, in the minds of most people, firmly established the existence of the Loch Ness creatures as an accepted fact.

In the meantime, the work at Loch Ness continues. One aim of the Academy of Applied Science is good, clear, color pictures of one or more of the creatures, on the surface, both movie and still. That the creatures do surface from time to time is accepted. Some witnesses even claim to have seen them on land, some distance back from the water, suggesting that they are amphibians. Dr. Rines went underwater for his first pictures and Dinsdale assisted him in his efforts. But Dinsdale's own efforts are directed toward surface photography and the modus operandi of the lone Britisher, when he is not assisting Dr. Rines, is a small boat loaded with camera equipment in which he quietly floats for hours, days, weeks at a time, patiently waiting. In 1975, after some thirteen years of patient watching and waiting, Dinsdale wrote, in answer to one of my letters,

"Thank you for your encouragement. Sometimes the water-hunt wears me down to the point where I feel like falling overboard on purpose. But then, like you, I believe that confrontation is coming. It's just a matter of sticking it out."

As mentioned, I have taken no part in the Loch Ness searches and when writing about them can only write of the work of others. But in the case of the Yeti, the snowmen abominable, I can contribute a little more. I was not only one of the first searchers for the elusive Himalayan primates but in the long run probably spent more time investigating that particular phenomenon than anyone else.

I cannot remember where or when I first heard of the snowmen. I seem to recall my father telling me bedside stories about them, years ago, and spinning yarns set in the swirling mists of the high Himalaya about the elusive hairy men that inhabited those inhospitable regions, stories, I realize now, that set a young boy's imagination on fire.

Many years later, my own personal experience with the phenomenon began, in conversations with Sherpas and Lepchas and Paharias in Darjeeling, the old British colonial hill town in northern India, during a one month, end of the war vacation I took from Royal Air Force duties (while waiting, with many others, to get on ships to go home) in 1946. What I learned then inspired me with a desire to learn more and eventually, I hoped, to take an expedition into the Himalaya and find the mysterious creatures.

Back in England, after demobilization, I started to do some research on the one-hundred-year-old mystery and one of the first things that I learned was that, apart from some mentions of primate creatures, in early Chinese manuscripts dating back to 200 B.C., creatures that might have been Yeti, there was really very little record before 1832. In that year the British Resident at the Court of Nepal in Kathmandu wrote in an article stating that some of his men, while on a collecting expedition for him, were frightened by a *rakshas*. The word *rakshas* is of Sanskrit origin and like the other name for the Yeti, shookpa, is still in use among the northern tribes of Nepal today. The Resident, Mr. B. H. Hodgson, an esteemed naturalist of his day, described the creature that his men had seen as walking erect, covered with long dark hair and tailless. Fifty-seven years later a British explorer, Major L. A. Waddell, found some sets of human like but unidentifiable footprints in the snows at 17,000 feet, in Sikkim, a neighboring country to Nepal also set in the Himalayan chain. His porters told him that the prints were those of a *shookpa*, or Yeti, and Waddell duly reported the fact on his return to London.

Another report came out of Sikkim in 1914. This time the origin was a British Forestry Officer and his find was also of footprints. J. R. P. Gent referred to the creature that made the prints as the *sogpa*, apparently a local, or lowland Nepalese pronunciation of the name *shookpa*.

The reports continued through the late 1800s and into the early 1900s. Lieutenant-Colonel C. K. Howard Bury, leader of the Mount Everest Reconnaissance Expedition of 1921, reported strange footprints on the Lhokpa Pass at an altitude of 20,000 feet. In 1925 a British photographer, one N. A. Tombazi, a gentleman of impeccable credentials and a fellow of the Royal Geographical Society, reported seeing a Yeti-like creature in the area of the Zemu Glacier at an altitude of 15,000 feet. The creature, he said, was moving through a stand of dwarf rhododendron bushes. (Twenty-three years later, in 1948, on one of my first expeditions, I found footprints in this same area.)

Then came Eric Shipton's find, a single, large, thirteen-inch footprint on the Menlung Glacier in 1951 and soon a host of famous names became associated with footprint finds, including my friend John (now Lord) Hunt, Sir Edmund Hillary, Frank Smythe and H. W. Tilman. The well-known anthropologist, Prince Peter of Greece, in whose former home in Kalimpong, India, I spent many pleasant hours, studied the question of the Yeti for many years and eventually concluded that there must be some basis in fact behind the many reports of sightings and footprint findings. The Scottish author, George Patterson and his lovely wife, Meg, (Margaret Patterson, F.R.C.S) who also lived in Kalimpong at that time, shared our interest and friendship.

In 1954 a British newspaper, the Daily Mail, sponsored the first Himalayan Yeti expedition. The party, which included my friends Gerald Russell and Ralph Izzard, spent many months of hard hiking and searching in the Himalaya without finding a Yeti. Then, three years later, in 1957, I came into the picture with the first American Yeti Expedition Reconnaissance.

My introduction to the magnificent mountains of the Himalaya began with the aforementioned trip to Darjeeling, in 1946, while still in the RAF., at which time I made what was hardly an expedition but more of a probe-a one week hike up the great ridge of Sandakphu, to about 15,000 feet, with a couple of Sherpas. Then in early 1947 I left the Royal Air Force and joined a British Tea Company, with offices in Eastcheap, London and tea estates in the Nagrakata district, in north Bengal. In those days, companies were liberal with the amount of vacation time they allowed their young gentlemen assistants and just a year later I was able to take a month's leave of absence and head for the mountains.

I left Nagrakata, in the central Dooars district, where I was based, by car-a little British seven horse power Austin Seven-and arrived in Gangtok, the capital of Sikkim, a day later. From there, with three Sherpas, one of them a sort of a cook, I set out on foot. My goal was the Lachen Lachung (villages) area, and the Green Lake region, in northern Sikkim, which I reached in some ten days of hard hiking. It was there, close to the Zemu Glacier, that I found the single footprint, in hard snow, at the edge of a small frozen pool. My Sherpas, two of whom were older men, were, I recall, very nervous about the finding and reluctant to stay in the area. We left a day later and returned by way of the Sikkim-Nepal border, along the eastern edge of the Hogla Bogla country of Nepal, traversing the Sandakphu ridge back to Darjeeling. We completed the last lap of the trek, some twenty-six miles, from Sandakphu to Darjeeling, dropping 10,000 feet down to 2000 feet and up again to 6000 feet, in one long day.

I was in Sikkim again in 1956. By this time I had resigned from my tea company and gone into the big game hunting business in Nepal. The life of a tea planter, wonderful in its heyday, proved less attractive after Indian independence and, with it, the end of the Raj and all that British colonialism offered a young white sahib in those great days, as well as which big game hunting safaris in Asia were in their infancy and just beginning to attract international attention.

In the spring of that year, with time to spare between safari bookings, I once again set out for Sikkim, again with the same three Sherpas but also with a pony, the daily trekking plan be to walk downhill on foot and ride up, or walk up holding on to the pony's tail. This time I chose a high ridge route running north out of Gangtok, one that started in the 6000 foot level and ran all the way to the Zemu Glacier in north central Sikkim. Once in the mountains I spent a month searching the dwarf scrub and frozen slopes up to 15,000 feet and found sign of wolves, snow leopards, Himalayan bear, Bharal (sheep) Thar (wild goat) and Musk deer, the common wildlife of the middle Himalaya. But no Yeti.

On my return, hiking hard with my three faithful Sherpas, I passed through the high moorland of Trejablo, west of Rathong, and one cold and foggy evening, marching over a high, rock-strewn ridge and looking for a way down to a lower and level camp site, we spotted a group of men trekking in single file, northwards, about 2000 feet below us. We wondered who they were and set out to cut them off and, if possible, share a camp with them, a month in the high mountains with three rough cut Sherpas having given me an appetite for some slightly more sophisticated human company.

Some three hours later we saw their campfire ahead of us and walking in out of the gloom I was delighted to receive a warm welcome from some old Darjeeling based Sherpa friends, including Ang Tharkey, Gyalzen Norbu, Ang Namgyal, Chuwang Sherpa and the now famous Tenzing Norgay, Hillary's immortal companion on Everest. After hot mugs of soup and an explanation of what they were doing there-they were part of the new Indian School of Mountaineering from Darjeeling-the talk naturally turned to the Yeti.

None of those present had ever seen one, but Tenzing told me that his father had encountered one which had crawled on top of his yak herder's hut one night and stayed there until driven off by smoke from the old man's yak dung fire. Tenzing also told me, again in regard to the Yeti, that he had recently talked with an American named Tom Slick, who was visiting Darjeeling and who was interested in projecting a major expedition to find one of the creatures. Tenzing's wife, he said, back in Darjeeling, had the American's address.

On returning from my second Sikkimese expedition, I met with Mrs. Tenzing in Darjeeling and she gave me a small, crumpled piece of paper which had Tom Slick's name and address on it...care of the National Bank of Commerce, San Antonio, Texas. I promptly wrote the man from Darjeeling-a registered letter, the only safe form of communication in India in those days-and some time later-the spring of 1957-met him in New Delhi, India, from where, after talks and the planning of a reconnaissance, we flew to Kathmandu together.

From Kathmandu, after some days of negotiation with Nepalese government officials, and enrollment of a group of Sherpas for our reconnaissance, we charted an old WW11 DC 3 and flew down to Biratnagar, in southeastern Nepal from where, accompanied by a Mr. N. D. Bachkheti, an Indian zoologist from New Delhi, we trekked up the Arun River valley and into the ten to fifteen thousand foot ranges of the northeastern Himalaya.

The reconnaissance, with Tom including its time of preparation in Darjeeling and Kathmandu, lasted from March 1st to April. We spent a month in the in the mountains together during which time I found one set of Yeti footprints, in the Chhoyang Khola at 10,000 feet, and Tom, working with a separate party in another area, found a second set.

The prints that I discovered were identical with those I had seen in Sikkim years earlier. They were ten inches in length, five-toed, and they were the prints of a bipedal creature of considerable weight. The prints that Tom found were thirteen inches in length and of similar construction.

We returned to Kathmandu where Tom, satisfied that a full scale probe was justified, immediately started making plans for a full-scale expedition to try and find one of the creatures and photograph it. Six months of preparation were put into the planning of the new expedition and in early 1958 we again set off for the mountains, this time with a party of five Europeans, ten Sherpa guide companions, and sixty-five porters. Our object, to find the Yeti, once and for all.

Our chosen area of search was the deep *kholas*, or valleys that drained into the upper Arun valley, including the Hongu, the Seeswa, the Aluwa, the Iswa, the Barun and the Chhoyang. Sponsors of the expedition were Tom Slick and Kirk Johnson, another wealthy Texan from Fort Worth and C.V. Wood, a Los Angeles industrialist. Leader in absentia would be Tom, unable to come on this phase of the search because of business and other commitments in the United States. Another field member would be Gerald Russell, a highly respected American naturalist who lived in France and who had already been in the Himalaya with the Daily Mail expedition. Expedition photographer would be George Holton, a professional photographer of international renown whose subjects were mainly in the field of natural history. Movies (16 MM) would be handled by a part-time movie photographer, German-born Norman Dyrhenfurth. Liaison with the government of Nepal, a required appointment in those days, was supplied by a Nepalese gentleman, Captain Pushkar Shumsher Jung Bahadur Rana, late of the Nepalese Army. And lastly I was to be in charge in the field-I think Field Leader was my title-with my brother Bryan as my assistant.

The 1958 expedition was named the Slick-Johnson Himalayan Yeti Expedition and a month after leaving Kathmandu, we established base camps high in the upper Arun. From these bases the party fanned out to cover different areas of interest and searches were combined with inquiries at Sherpa villages on possible areas of habitat. Once every two weeks all of the members of the group came together and compared notes.

After three months only one possible sighting had been made, by a Sherpa who was a member of the guide party, named Da Temba. Gerald Russell was with Da Temba at the time of the sighting but did not see anything. However he believed that the man did see a Yeti and later we spent considerable time in the area of the sighting. Then, one by one, the various members had to leave and before four months had passed all, with the exception of my brother and I, had departed.

We stayed on for another five months, making nine months in all and it was not until the snow was deep on the ground that we marched out to Kathmandu, a cold, two-hundred-mile trek that used up the last of our stores and wore out the last of our equipment.

From Kathmandu, in late 1958, we corresponded with Tom Slick. We had sent monthly reports to both him and Kirk Johnson all during the first nine-month expedition and now it was a matter of assessing the value of returning to the Himalaya and continuing the search. In the meantime we were to rest in Kathmandu and this we did at the old, now abandoned, Royal Hotel, a hostelry of comparative comfort for the weary Yeti hunter just down from the mountains.

There we stayed for some six weeks, until December, 1958, when we received our second set of orders. They were simple. We were to return to the mountains and continue the search. We were to stay up for as long as it was necessary to find and photograph a Yeti or, failing that, for as long as we could stand the cold, the privation and the various discomforts of living under rugged conditions at high altitudes.

We set out again a week before Christmas and if the reader wonders why two young men would leave the delights of a Kathmandu hotel and the fleshpots of an eastern city and its swinging international set a week before Christmas, all that I can answer is that in those days we were very keen and very dedicated to a task to which we were prepared to devote years, if necessary, to what Tom Slick called the Ultimate Quest. It might help to show how keen we were, perhaps, if I were to add that for that first year's work with the San Antonio group we accepted no salary and for the second, for needed personal expenses outside of the expedition proper, we worked for an honorarium of $100 per month each.

For the second expedition-or the third, if the reconnaissance could counted as an expedition-my brother and I went alone with a small group of Sherpas. We took a minimum of equipment with us. No tents, no food. We planned to live off the land and to sleep, as the high mountain shepherds did, in wood shelters when we were below the tree line and in caves when we were above it. This we did, for another nine months, and they were nine very fascinating months during which we believe that we made some contribution to support the theory that there were still a few of the mysterious primates living in the deep, heavily wooded, seldom penetrated valleys of northeastern Nepal.

Among our finds were the now famous Pangboche temple scalp and the mummified hand of what might have been a Yeti, or a man, a question which has never been satisfactorily determined. The scalp that we unearthed at the Pangboche temple is still there to this day. The scalp that Sir Edmund Hillary's group brought back to the United States and which with much fanfare was subsequently found to be nothing more than a fake, made of the skin of a Serow, a species of Himalayan goat antelope, was simply a copy of the Pangboche scalp, made by a visiting Tibetan taxidermist many years before, to ease an inter-temple rivalry that existed in the matter of Yeti scalp possession between the temples and monasteries of Pangboche, Thyangboche and Kumjung. As to the hand, after a generous donation to the lama custodians of the temple, we managed to take part of it, one finger, the forefinger, out of Pangboche and have it sent back to England for examination. The task of getting it out of Nepal was simple; I hiked across the border with it, in a backpack. The undertaking of getting it out of India, by air, was more difficult. But this challenge was solved by friends of the Kirk Johnsons, who were passing through Calcutta after a visit with an old friend of mine, the Maharaja of Baroda. Via coded cables, Tom arranged for me to meet with them at the Grand Hotel in Calcutta … the famous actor, Jimmie Stewart and his wife Gloria. They, much amused at the thought of the grisly trophy that they were carrying with them, took it back to London with them and gave it to our resident "*boffin*" there, the renowned paleontologist, Dr. Osman Hill, for examination.

We carried the search through all of 1959, trekking and camping in canvas tents and spending much time in caves-where, as opposed to a tent, one can sit at a fire and keep warm. We hiked more than five thousand miles across the high Himalaya, living off the land-lots of potatoes and *champa* and Tibetan tea-enjoying the company, from time to time, of the mountain people, Nepalese, Bhutias and Sherpas and experiencing and enjoying a life as rough and adventurous and at times as dangerous as any young man could possibly desire.

CHAPTER SIX
The Bigfoot Mystery and its Shadowy Beginnings

The whole history of scientific advance is full of scientists investigating phenomena that the Establishment did not think were there. MARGARET MEAD.

The search for Bigfoot began again for me in the winter of 1959. At that time my brother Bryan and I were winding down the last of our three long Himalayan Abominable Snowman expeditions. In these last winter months, with deep snow on the ground and temperatures to 45° below zero, we were living in a cave in the upper Choyang Khola, one of the deep river gorges that runs down into the headwaters of the mighty Arun River, in northeast Nepal.

We were tired. Most of our equipment had long since been destroyed, or lost, or simply worn out. Our tents were used to the point of no return and then dumped. Our clothes were in rags and we were subsisting on *champa*-a Sherpa grain mash not unlike oatmeal-yak's milk, yak milk cheese, potatoes when we could find them, edible ferns and edible nettles.

One bitter December evening, while we were sitting at our fire in the cave, our Sherpa mail runner, Ang Namgyal, arrived from Dharan, the British Army Gurkha soldier recruiting depot in southern Nepal, where a Major Allen kindly handled our mail for us. The runner, who had completed the 140 mile hike in four days, came pounding in out of the dripping gloom of the Himalayan night and handed us a package of letters in a leather pouch. One of the letters was from Tom Slick.

Tom's letter was interesting. It contained instructions for us to terminate our three-year hunt for the Snowmen and close down the expedition. We were to proceed to Kathmandu, the capital of Nepal, where we would find further instructions awaiting us. In the time that it took us to march out, the letter continued, we were to give consideration to coming to the United States to take over a new expedition that would search for something called the Bigfoot, a giant primate not unlike the Yeti. It, or they, apparently lived in the Pacific Northwest of the United States and their habitat was the thickly wooded coastal ranges of the northern states of Oregon and Washington. Before coming- if we decided to take up his offer-we could take a break, as long as we wished and when ready inform Tom, who would arrange appropriate travel plans to the U.S.

The letter brought to an end our Himalayan odyssey and knowing that we might never again see or visit many of the places where we had spent close to thirty six extraordinary months, it was with mixed feelings that we closed down the expedition. It took us a couple of days to pay off our local Sherpas, dump or give to them worn out equipment that we no longer would need and begin the 200 mile trek to Kathmandu. We arrived in Kathmandu eight days later, December 30th, checked into the Royal Hotel and, again on Tom's instructions by mail, made plans to meet with him in Delhi, in India, a week later.

The series of events that took place after this included a meeting with Tom in Delhi, a road trip down across India with him to investigate -as part of his interest in what he called "mind science"-native claims of strange and unusual capabilities, a trip back to Ireland for Bryan and a separate one for me to Australia to visit my parents, who had recently moved there. So it was several months later, May, in fact, before I was able to get to the U.S., meet with Tom in Texas and get to northern California to open up the first Bigfoot project, later to become known as the Bigfoot Research Project # 1.

Late May found me temporarily-pending permanent quarters-installed in a small motel in a little town called Willow Creek, in Humboldt

County, in northern California. There, a stranger in a strange land, I set about trying to organize the first Bigfoot expedition and make a systematic and sensible approach to the problems of finding out, A) if the things really existed, and, B) if so, how to go about finding one. To start, however, I had to learn something about the people I was going to be dealing with, by whom I mean local persons of the area where we were going to be based. As I soon found out, they were a diverse lot, ranging from the serious and intelligent and dependable to characters who could best be described as colorful, among the latter a trio of gentlemen who, until I realized what kind of jokers they really were, occupied not a little of my time.

The three in question were engaged in the logging and timber trades and when they approached me it was via the one who, he said, was their leader and spokesperson and who, as such, would be in charge of any negotiations between us, these to be an exchange of Bigfoot "evidence" for which they would expect to be paid. As the ensuing farce was something that I can never forget, I feel that it deserves telling in some detail. A few days after they first they contacted me I met their so-called leader, by invitation, at his home outside of Willow Creek.

I never learned the names of the other two men of the trio. But their leader was a man named Ray Wallace-a name that later became quite well known in the world of Bigfoot for its association with the more comical side of the phenomenon-and it was he that I met soon after they first contacted me.

Wallace was a tall, angular man of about sixty, with a hard, thin face, a large, bony nose, close-cropped, greying hair and long yellow teeth. Perhaps because of his work amid the constant roar of high-powered logging machinery, he had acquired the habit of shouting. Thus he never spoke, but shouted all the time. Yes! He shouted, ushering me into the house, I had come at exactly the right time. His companions, he roared, his "range and track teams," as he called them, were real woodsmen and at this moment were hot on the trail of a goddam Bigfoot. They expected to capture the bugger at any moment. They were first-class trackers and top woodsmen and they were doing something that none of these yahoos who worked for that Texas fellow were able to do. Hell, man, these guys could track a squirrel up a tree and down again. In a few days he would be calling me and I could rest assured that he would have something to show me then. He shouted me to the door, shook my hand heartedly and I left with my ears ringing.

Sure enough, three days later, he called. The phone crackled in my ear. We have it, he roared. My range and track team has done it! We've got the bugger! It's only a young one, but it's a real goddam Bigfoot. Why shit, man, it's got hair all over it and the biggest goddam feet you ever did see. You wait until them sciences fellows see this bugger.

When I could get a word in, I asked him when I could see the beast. Why man, he replied, as soon as them sponsors of yours come up with our price, that's when. I gently asked him what the price might be? One million dollars, in cash, and they would hand the beastie over, complete with carrying cage.

I called Tom in San Antonio and told him about the situation. He told me to offer Wallace five thousand and to tell him that we would pay it on sight.

I called Wallace and told him what the offer was. At first he would not hear of it. Five thousand, he bellowed. Not a chance in hell. But at the same time he would call his partners and talk to them. And then get back to me. He would call me in a couple of days.

Two days later he called and said that they would accept the five thousand but that they would have to have it in advance. Nothing doing, I told him and from there we went into two weeks of phone calls and argument back and forth, while we bargained for a look at the creature and they bargained for cash on the nail, sight unseen. Then came an urgent call from the man. He and his partners were getting into difficulties. The only thing that the young Bigfoot would eat was Kellogg's Frosted Flakes and it ate them by the hundred-pound bag. The cost was running them dry financially and I would have to do something quickly. There was only so much money available for Frosted Flakes.

I called Tom again and told him about the Frosted Flakes problem. He told me to offer them $500 for a look at the creature and to take a camera along. I did this; they said no.

About this time Tom has scheduled a visit to see how things were going and how we were moving along with setting up the base. He told me when he was coming in and asked me to arrange a personal meeting with Wallace and soon after he arrived the three of us met in a café in Willow Creek.

Like me, Tom was highly skeptical of Wallace's claim. Nevertheless, he decided to play along with the man and after some brief opening conversation, we got down to business. Pulling out a check book, Tom wrote a check for $25,000, made it out to Mr. Ray Wallace and put it on the table in front of the man. This, he said, is what I am willing to pay because if you do have what you say you have, it will certainly be of enormous scientific value.

Wallace, his eyes bulging with surprise and delight, reached for the check. But as he did so, Tom's hand, which had never quite left the little piece of paper, gently drew it back. "It's all yours, Mr. Wallace," he said. "It's made out to you and there are ample funds to cover it. But, if you don't mind, we would, er, well, you know, just like to have a look at the creature before you cash it?

Wallace left the cafe, saying that he would have to consult with his partners but agreeing to meet with us there again the following morning.

We met again next day. Same time same place. But this time Wallace, instead of the hale and hearty and shouting fellow he had been the day before, arrived with a look of woe on his face and sat down to join us with a long deep sigh.

As Tom and I suspected, Wallace's "poor little Bigfoot" got sick in the night. Dern thing started throwing up. Got kinda pale in the face you now, and has us real worried. So we let it go. I mean, that's what you would have done, right?

Right. We did not hear from Mr. Wallace again for a long time. The question remained of course, would he have cashed the check if Tom had given it to him? I personally don't think so. I met him again later and found him to be a decent enough man. A joker, yes, but not a hoaxer, or a con man, as were some of the characters the phenomenon see med to attract.

The expedition team I eventually put together consisted of my brother Bryan, Steve Matthes, a professional U.S. Fish and Wildlife lion hunter from California, and myself. We were joined from time to time by Tom Slick, always bringing with him interesting people from the United States and other countries. We found a total of five sets of footprints during that year (1960). We did not see a Bigfoot. But we worked hard at the problems of finding one, and many good people helped us in our work honest, dedicated people who were willing to give of their time and their experience for little or no return. The townspeople of Willow Creek and Salyer cooperated wholeheartedly in our efforts and generally tolerated the idiosyncrasies of the two foreigners who spent all of their time looking for something which, to many of the them, probably did not even exist.

Our efforts were terminated and the project came to an end with the death of Tom Slick in the crash of a small private plane, in October, 1962. Tom and his pilot died in a small two-seater that broke up in a storm over Dillon, Montana. His body was flown to Texas for burial and with his demise the first Bigfoot project came to an end. We looked upon Tom Slick as a man of vision and enterprise, a fine and gentle man, always sensitive to the feelings and needs of others. His death was a great loss to us all. Since those first Himalayan days together, he had been my friend.

I returned to Nepal and resumed my big game hunting and guiding career which had been interrupted by the Yeti and Bigfoot expeditions, and I stayed there, running Nepal Safaris, my own professional guiding company, until early 1968. In 1968 I gave up hunting and traveled to Washington, D.C., where, with four other Washingtonians-Karl Jonas, M.D., Roy Lyman Sexton, M.D., and two attorneys, Leonard A. Fink and Scott C. Whitney-I founded the International Wildlife Conservation Society, Inc. In late 1968 I returned to Nepal and, working for a year with the government of that country, created the Sukila Phanta (White Grass Plains) Wildlife Reserve, the first tiger sanctuary in Asia, a 200,000 acre preserve that today offers one of the last retreats for the great cats in Asia.

I completed this work in late 1970 and returned to the United States, where I found, after eight years, that there was renewed interest in the Bigfoot phenomenon. I made some inquiries and found that some backing was available if I was prepared to put up a fair proportion of the financing myself and so in early January, 1971, I and a few companions started another search.

We chose as our base the little settlement (half a dozen houses) of Evans, Washington, just south of the Canadian border, and we worked from there for one year.

In early 1972 we moved to The Dalles, a township of some 10,000 people, in northern Oregon on the Columbia River. At the time of writing, four years has elapsed of the new search and investigation and a permanent base has been established in The Dalles, Oregon, which includes my own home, an equipment storage facility, and an Exhibition and Information Center. The latter is in a 45 foot mobile home and is permanently housed on a site at West Sixth Street in The Dalles. The Exhibition is self-supporting and also helps to make extra income for the search and investigation project. But its real value is as a conduit for information about the phenomenon. Opened in May of 1974, it filled a need in the Bigfoot field. Once it became established-with the help of The Dalles Chamber of Commerce and the excellent publicity that it obtained in newspapers, radio, and television-people began to come in with information about the phenomenon. Much of it was old or too insubstantial to be of any real value. Some of it was even imaginary. But a great deal of it was very useful, especially where subsequent investigation showed the people who gave it to us to be men and women of character and integrity. This information, on sightings and on footprint findings, became the foundation on which we have built most of our search and investigation plans. It also helped us to begin, at long last, to form the Geo Time Patterns that we felt were essential to the search and which in time could lead to success. The main reason for this excellent flow of information was that people soon begun to realize that the Information Center was a place where they could tell their stories; a place where they would receive a serious reception, no matter what they had to say; a place where they would not be laughed at or ridiculed in any way; and where their names and their stories would be held in the strictest confidence.

Four years of work in the new search have produced an extraordinary amount of new evidence. At the Information Center the files bulge with information on sightings, footprint findings, migration routes, eating habits, food, coloration, size. Hundreds of letters have been received from people all over the United States, and with the establishment of our monthly newsletter, THE BIGFOOT NEWS, in September 1974, this flow doubled and then tripled. Visitors came from all over the world; some even stayed to work on a volunteer basis. Of course they asked a lot of questions and one of these, that surfaced quite often, was in regard to remains, meaning carcasses and bones and why none were ever found. How was it there are no skeletons, no skulls, no bits and pieces found lying around?

There are four answers. A) They do not exist. B) They bury their dead and bury them deep. C) Their bones dissolve and disappear. D) Something comes along and eats them.

Let us deal first with A) They do not exist. If we believe this then there is no further argument. No bodies, no bones and that is all there is to it. But if they do exist? That takes us to B) They bury their dead. This is not an unreasonable assumption. Some of the Neanderthal men buried their dead, as we have fairly recently discovered, and the Neanderthals were a decidedly primitive lot. When he was working with Roger Patterson and headquartered at Yakima, Dennis Jenson saw a letter from a man who swore that he had watched three Bigfeet burying a fourth. They dug a deep hole, using only their hands as tools. After placing the body in the hole and covering it with earth they rolled huge boulders, each weighing many hundreds of pounds, onto the grave. The letter, with its extraordinary story, was lost from Patterson's files after he died in January 1972. Jenson, who saw the letter only once, was unable to remember the name and address of the man who wrote it. If the letter was a true account of an actual burying, then the Bigfeet do bury their dead. If they do, this would help to explain why we do not find any skeletons. If they do not then we have to deal with C) Their bones dissolve and disappear. A nice little theory and one that conveniently disposes of the "no bone" problem. Could it be true? The answer is yes. It depends almost entirely on soil conditions.

Fresh bone is composed of organic protein fibers. When a bone begins to fossilize, the protein material disappears. Percolated mineral material, usually from ground waters in the surrounding soil, takes its place. This is followed by various chemical reactions at the molecular level and the bone, through the course of time, slowly changes to stone. The chemical changes are such that in pure fossilization, the shape and size of the bone remain unchanged. All this depends on soil conditions.

Bones fossilize in wet alkaline soils and in dry alkaline soils. In the latter, light sub-fossils are usually found. In wet soils that are airless, such as the peat bogs of Denmark or the turf fields of Ireland, complete preservation may occur. But in wet acid soils there is no fossilization and the bones disappear. The soil composition of much of the mountain ranges of the Pacific Northwest, our supposed area of Bigfoot habitat, is wet acid.

And lastly, D) Something comes along and eats the bones. Another convenient answer? No, for the forested ranges of the Pacific Northwest have a natural disposal system that is as efficient, if not as violent and quick, as the African and Asian systems. It consists of a family of carrion eaters that includes crows and ravens, buzzards and other meat-eating birds, including eagles. The eagle, incidentally, proud predator though he seems to be, will not hesitate to eat carrion when there is a free meal around. Other members of the disposal squad are coyotes, wolves and foxes, various rodents, and of course, porcupines.

And the leader of the pack? Euarctos, the Black bear. Led by the Black, with his extremely keen sense of smell, this highly efficient group of carrion disposers works tirelessly at its job of keeping the forest clean. Everything is eaten, and if the particular carcass happens to have antlers, these are eaten by the horn and hoof and ivory-eaters of Asia and Africa, the porcupines.

In the course of the my investigations I have questioned many people about finding bones; I have never yet met anyone who has found the bones of a cougar or mountain lion, only one man who has found the bones of a dead bear, and perhaps half a dozen who have seen complete deer skeletons. These latter, when found, were usually quite fresh.

The second most common question, again often asked as a challenge, and usually from a fairly skeptical source, is this. If the Bigfeet exist, what do they eat in the wilderness? Surely there is hardly enough to support the dietary needs of a creature as big as it is supposed to be? For instance, what would it eat in the winter?

The answer to this one is actually fairly easy. There is, in the wilderness areas of the northwest and also in British Columbia, an abundance of food for the omnivorous creature we would suppose the Bigfoot to be. And I know whereof I speak, for on several occasions I have been forced to fall back on natural foods when my supplies ran out, and I have been able to keep the inner man happy every day without undue effort.

Creatures of the size that the Bigfeet are supposed to be would have to be omnivorous to survive. That is to say, they would have to eat everything that was available. Their diet would therefore include, but not be restricted to, small animals and fish, water life of all kinds, including frogs, prawns, crabs, hermit crabs, newts, water lizards, water snakes, various crustaceans, limpets, cockles, mussels and clams. Like the bear, they would eat carrion of all kinds, and it is possible that their sense of smell would be developed to the point where, like the bear, they would be able to scent decaying carcasses from a great distance and then locate them by scent. They would eat birds' eggs and, of course, young birds.

In the herbivorous area, the diet would include bark and the under bark of certain trees which is edible even to man, sedges, lake weeds, berries, rose hips, fruit, nuts, shoots of certain edible bushes, many forms of grass, leaves, flowers and many other kinds of wild plants.

In the insectivorous area, all kinds of insects, including moths, butterflies, bees and their honey, worms, ants and their larvae, and beetles.

Like the bears, they might also be cannibalistic and this of course would be another reason why we just do not find any of their carcasses lying around in the woods. (A male bear, for instance, will, given the opportunity by a careless mother, kill and eat his own young.)

The third most oft-repeated question concerns habitat and space. The people who ask it are, more often than not, city dwellers, or foreigners from places like Chicago or New York, who have little or no conception of the size of the Pacific Northwest ranges or the extent of the vast reaches of forest that cover the mountains. The question is, where could the things live? Where could they hide? How is it that more of them are not seen with the mountains full of fishermen and hikers and snowmobilers?

In 1972 the media launched upon a startled world the story of a lost tribe, the Tasaday, found deep in the jungles of the Philippines. The newspapers, their interest in lost tribes and lost men stirred by the finding of the Tasaday, also speculated on the number of Japanese soldiers who might still be hiding in south Pacific island jungles. Over the years, as we know, several Japanese soldiers have come out of hiding. Sure enough, during that year, another Japanese appeared, twenty-five years after the end of hostilities. The Tasaday, of course, were not hiding. They were just there, waiting to be discovered. There were not very many of them, true, and they did live in dense forest in a little-penetrated area. But they really did not mind being discovered and might even have wanted to be found, if they had known that there was anyone to find them. The Japanese soldier, on the other hand, was in hiding, as he thought, from American soldiers, and up to the very last moment he made every effort to hide and remain unseen. The area of the Philippines, where the Tasaday were found, is 115,707 square miles. The total area of Guam, from the jungles of which the valiant and still-fighting Japanese soldier appeared, is 209 square miles. And the area where he-and until recently several of his companions had managed to remain hidden for some twenty-eight years, has a human population of 87,000 people.

What is known as "the area of habitat" of the Bigfoot are those regions which show credible evidence in the form of history, footprints, and sightings. It comprises part of northern California and part of northern Idaho. It includes all of the coast range and all of the Cascade range of Oregon and Washington and all of the coast range of British Columbia, north to a line drawn east-west just below Prince Rupert, in northern British Columbia. The size of this area, the actual area that has produced and that continues to produce the evidence of which I write, is not less than 125,000 square miles.

This vast area is composed, for the most part, of very rugged country. It contains many high mountains that hold permanent snow on their peaks. Visible from The Dalles, Oregon, for instance, are three peaks, all in excess of 9500 feet. There are deep valleys and gorges with ice cold streams, high ridges and cliffs and hundreds of miles of dense forest. There are many government-designated wilderness areas, where, even if there are roads, no wheeled transport of any kind is allowed. There are dozens of deep ravines that never hear the voice of man. For the most part the total area is unoccupied by people, for people in the United States and in Canada seem to prefer to live in the lowlands, where life is less harsh and where the amenities of civilization are within reach. British Columbia itself has a total area of 366,255 square miles and a population of only 2, 190,000 people. Three-quarters of this population live in one-fifth of the total land area. The remainder of British Columbia is very thinly populated.

Again, the Bigfeet, it is obvious, are obviously not sitting around waiting to be discovered. Unlike the Tasaday, who lived in a very small area of jungle and made their permanent homes in a series of caves, the Bigfeet, the evidence suggests, are constantly on the move and, like a tribe of Australian aborigines gone "walkabout," are here today and very much gone tomorrow. Also, unlike the Tasaday, and much more like the aforementioned Japanese soldiers, they do not want to be found and with a range as vast as their habitat suggests, this should really not be too difficult.

Statistics show, in fact, how well protected they are by the nature of their habitat, its cover, and its size. A quick glance at the Bigfoot search guide should clarify this for anyone who has ideas about finding one, either alone or with a small army of men. The chart statistics are quite serious, even if the idea of a stationary Bigfoot is not.

These questions of fossil bones, food, and habitat cover will, no doubt, be discussed again and again. The skeptics, annoyed that I should put forward such puny answers to their challenges, will take my theories and hurl them out of the nearest window. Alas, after five years of dedicated research on the subject, they are the best that I can do at this time. Perhaps in time I shall be able to do better. With determination and enterprise even a mouse trap can be improved. Or so people tell me.

CHAPTER SEVEN

The 1967 Bigfoot Footage, a Big Man in a Fur Coat? or The Real McCoy?

If it is a fake then it is a masterpiece and as far as we are concerned the only place in the world where a simulation of that quality could be created would be here, at Disney Studios, and this footage was not made here. Statement to the author by the Chief Technician, Disney Studios, Burbank, California, January, 1973, after an examination of the 1967 Bigfoot footage.

NOTE: The following account of a safari in the country of Nepal, written as a preliminary to this chapter (and its examination of the 1967 footage,) was included in the original manuscript for the sole purpose of establishing the fact that I was not in the U.S. on the date when the footage was shot and, because of this, had no association of any kind with the footage and the people involved.

My safari journal for October 20th 1967 tells me that on that day I was deep in the jungles of southwest Nepal, guiding a very difficult, rather gnome-like gentleman from Nashville, Tennessee, on a leopard hunt. Colonel Landan Daniels was 65 years old, very wealthy and permanently addicted to sour mash whiskey and cigarettes. He had a hacking cough, he talked incessantly, and he was one of the worst shots that I had encountered in two decades of big game hunting. I had refused to allow him to sit in a tree to shoot his leopard. It was simply against my principles and I told him that if he wanted a leopard trophy he would have to shoot it on the ground.

We had been stalking the leopard, on foot, for two days. Now, October 20th, was the morning of the third day and I had a fairly good idea on the big cat's movements. I had carefully instructed the colonel. I had told him that until we got close enough to shoot, all that he had to do was follow me, very carefully and very, very quietly. When the time came to shoot, I would tell him. In the meantime he was to look only at the ground, at where he was putting his feet, and he was to put his feet where he was actually looking and not somewhere else. Silence, absolute silence, was essential. The slightest noise and the leopard would detect us and be gone in a flash. The breaking of a tiny twig, the crackling of a dry leaf underfoot, would be enough to have the big cat perform its disappearing trick. Once that happened we would have to start looking for another leopard.

The morning of October 20th had dawned bright and clear. The mild, dry, winter weather of the southern Nepal jungles made for good visibility and great hunting. Now it was about 7 AM and after two hours of stalking we were about to close in on the big cat. We moved very slowly now. Every inch of the ground was carefully studied, the pattern of the pug marks was analyzed and the jungle ahead and on both sides was watched with the greatest of care. Behind me, breathing rather heavily and occasionally making vague choking noises as he tried to smother his smoker's cough, moved the colonel. Excitement rose as we moved closer, and I could hear the colonel's breathing getting faster. Suddenly a group of Rhesus Macaques monkeys began to chatter over our heads. They had seen the leopard, somewhere ahead of us, and they were warning the other jungle folk of the danger. Instantly the leopard uttered a deep growl of annoyance, a sound that enabled me to pinpoint its position.

I moved to the right a few feet and the colonel did the same behind me. Ahead of us the jungle opened up; there was a small glade with short grass and very light undergrowth, its green leaves still wet from the night dew and now shining in the morning sun. The colonel, his breath now beginning to sound like the puffing if a distant steam train, moved in behind me and we had both stopped to watch and listen when the leopard walked out into the glade and turned and looked directly at us.

I remained perfectly still and, keeping my eyes on the big cat, whispered, over my shoulder and in the lightest of whispers, to the colonel, telling him to shoot. For a few seconds there was silence. The monkeys, sensing a jungle drama about to be enacted, stopped chattering. The colonel stopped breathing. There was no sound and the big dog leopard, his black-rosetted coat of orange magnificent in the morning sun, remained frozen in position. For a moment it seemed as though time had stopped. Then the colonel made his play. He leaned forward and took a deep breath. He pointed a crooked and shaking forefinger over my shoulder and in a voice that shattered the jungle silence, set the monkeys to screaming and leaping for their lives and alerted every animal within a mile, roared, "That's a leopard!" Needless to say, we never did see that leopard again.

We returned to camp and had breakfast and talked about different tactics and so passed another jungle day, routine, ordinary and, except for the humor of the leopard incident, uneventful. But for two men in the woods of northern California, some twelve thousand five hundred miles away, it was a far from ordinary day. For, if their story is true, in the afternoon of this day, October 20, 1967, they encountered and were able to photograph, for the first time in history, on movie film, a large female Bigfoot.

The film, or rather the footage, which is what it really is, consists of twenty-eight feet of 16mm color movie footage taken with a hand-held camera. The footage is of poor quality and, with the exception of a few frames at the beginning of the sequence, in the middle, and at the end, which are sharp enough to show detail, the footage is blurred and unclear.

The subject of the footage is a large, hairy, upright-walking figure. It is seen walking from left to right, turning once to look in the direction of the cameraman and then disappearing into the trees. The site at which the footage was made is the partially dry stream bed of a stream called Bluff Creek, a stream that begins in Del Norte County, flows into Humboldt County and empties into the Klamath River; these counties are in northern California.

The photographer and his partner report that the time at which the footage was obtained was somewhere between 1:15 P.M. and 1:30 P.M. and the date, October 20, 1967.

The 1967 Bluff Creek footage has been and is the subject of much controversy. Since the rights were purchased from the original photographer by a commercial company, which then distributed it throughout the United States, very many people have seen it. It has also been released in Europe and Australia and other countries and it has been seen on television as well as in movie theaters by thousands of people.

 The question is, of course, is it real? Is the subject of the film a real living creature-a large unclassified primate- or is it a man dressed in a fur suit? Here, in this chapter, I am going to set down on paper all that I know about the film and the filmmakers. I will add to this some of the latest findings on the film in reports from groups of expert viewers and examiners in three different countries. I feel competent at least to discuss the footage, for I have spent many hours with the two men who obtained it, have viewed the footage, for purposes of examination, perhaps a hundred times and have visited and studied the site where it was made in Del Norte County, California.

The two men who obtained the footage were from Washington state. Their names were Bob Gimlin of Union Gap, and Roger Patterson, of Tampico. Union Gap lies in the Yakima Valley and through time has become almost an extension of the city of Yakima. Tampico is a scattering of houses and small farms about twenty miles west of Yakima.

Bob Gimlin is a horse breeder and cattleman. Roger Patterson, who died in January, 1972, was a horse breeder and rodeo rider.

Both became interested in the Bigfoot search in 1963 and from that time conducted, when time out from work allowed, their own private searches in Washington, Oregon, and northern California. Sometime in these years Patterson leased a 16mm movie camera and began carrying it with him. They had agreed that if they could obtain footage of a Bigfoot, they would not attempt to shoot one. As a precaution, however, they both carried rifles.

Being horsemen, their method of hunting was from horseback and when they went out on a search, their equipment consisted of two riding horses, one pack horse and a large horse-carrier truck.

In October, 1967, they took their horses and drove down to Bluff Creek in northern California-the area where we operated in the sixties. It was, even to 1967, an area that consistently produced what was believed to be Bigfoot footprints. Arriving there, they made camp on the west bank of Bluff Creek, just below where it is joined by Notice Creek and from there commenced their search. The bed of Bluff Creek, in its upper reaches carries very little water. The stream, some six inches deep, occupies only a few feet of the actually bed. The stream bed itself, between its permanent banks, is probably one hundred yards wide.

In the afternoon of the day of the incident, the two men were riding north in the bed of the creek. Both men were mounted on their own horses and behind them, trailing on a rope, came a pack horse. The pack horse carried light camping equipment and some food and was their safeguard against being unable to get back to camp in the same day. It was a fine day, with sunshine and no wind; the hooves of the horses made little or no noise in the soft grey sand of the creek bed.

After about half an hour's riding they found themselves approaching a big, jumbled pile of logs lying almost in the center of the stream bed. The logs, some of them very large and still carrying their roots, were the result of a sudden flood the previous year that had washed them out of the higher reaches of the creek and piled them up at this point. The pile, or logjam, was probably one hundred feet in width and fifteen feet high. It effectively hid the approach of the horsemen from anything that might be behind it.

Both horsemen rode around the logjam together-Patterson in front and Gimlin trailing the pack horse behind him-and, according to their story, immediately saw a large, female Bigfoot squatting on the right (or west) bank of the watercourse ahead and to their left.

Patterson said that she was holding a large black rock in her hands and had it raised to her mouth. Seeing the two men, and the three horses, she immediately stood up, turned to her left and started to walk away. She took two, maybe three steps, stopped, stood looking straight ahead (not at the riders) for about three seconds and then commenced a steady, purposeful, arm-swinging stride which took her upstream, across a sand bar on the right (west) side of the stream and eventually carried her into a big clump of streambed trees and out of sight.

Things happened very quickly after that. All three horses panicked and both men were, again according to their story, shocked and alarmed at the sight of the huge creature. The pack horse reared, broke its trailing rope and bolted downstream. Gimlin's horse began to buck in panic and he was forced to dismount in a hurry. He slid out of the saddle and took a firm grip on the reins and managed to hold the horse. Patterson had less luck. His horse reared and then fell over sideways, coming down on his right leg, crushing the metal stirrup on his foot and pinning him temporarily to the ground.

While Patterson was struggling to get up, the Bigfoot was walking rapidly away. Then Patterson's horse struggled to its feet and he rose with it, pulling the 16 mm camera out of his saddle bag and shouting to Gimlin to cover him with a rifle as he ran after the Bigfoot. Like the pack horse, his horse also bolted.

Patterson aimed his camera at the creature and kept his finger pressed on the trigger while he changed position three times and until he ran out of film. There was, according to him, only twenty-eight feet of film in the camera. The remainder had been used earlier to shoot other subjects that were to be part of a documentary film on the Bigfoot. Then, as suddenly as it had started, the camera was empty, the Bigfoot was gone, and it was all over.

The two men rejoined each other and, after a brief discussion, decided that it was more important to go back and look for the horses-both the pack horse and Patterson's horse, which had also bolted as soon as he left it-than to pursue the Bigfoot on foot. This they did, finding the two horses grazing quietly a mile downstream. They retrieved the horses, returned to the scene of the photography, tied up the horses and set out to try and track the Bigfoot. This they were unable to do and so after some searching they came back again to the photography site, made plaster casts of the footprints that the creature had left in the soft sand and then, with the horses, returned to their base camp near Notice Creek.

They closed up the camp and, leaving the horses tied, with fodder, took the horse truck and drove out to the coast, to send the film back to Yakima for processing. Taking a short cut to the coast, they were able to reach a small plane airport called Murray Field-located between Arcata and Eureka-before dark. There, according to Patterson, they left the film for dispatch to associates in Yakima, by private plane, the following morning.

From Murray Field they returned to their Bluff Creek camp, but this time via a different route, through the town of Willow Creek, where they stopped to visit with a friend, a Mr. Al Hodgson, the owner of the Willow Creek Variety Store. From there they drove on, back to their camp in Bluff Creek from where, after loading the horses, they exited the area and drove nonstop to Yakima.

They reached Yakima the following day, the 21st and on the 22nd gathered with their associates and families to view the footage, which had been quickly processed.

The 1967 footage, as I mentioned before, caused a great deal of controversy. Opinion, among those who have seen it, is much divided. The burning question, of course, is the reality of the subject of the film.

American scientific opinion on the film is, for the most part, fairly blunt. As far as American scientists are concerned, the film is a hoax. The subject of the film is a man in a rather cleverly made fur suit and that is all there is to it. But the trouble with scientific opinion is that with only two exceptions that I know of, not one single scientist in this country has taken the time to inspect the film, make a careful examination of it and/or subject it to expert analysis. The film has been seen by primatologists, mammologists, biologists and zoologists at the Smithsonian, the Natural History Museum of New York, the Primate Institute at Emory University and the Primate Institute at Beaverton, Oregon, the latter probably the foremost institute of its kind in the country. The "examination" in each case was usually limited to one viewing, after which the film was simply declared a fake. I myself have been present at some of these viewings and the reader may be interested in seeing some of the weighty scientific opinions that were directed at the film. I will quote three of them, as follows:

1. Nothing walks like that. Therefore it is a fake.

2. The subject in the film has breasts that are hairy. Apes do not have breasts and humans do not have hairy breasts. Therefore it is a hoax.

3. If they were there, we would have known about them by now. Therefore it is fabricated.

Two men, both eminent in their respective fields, have had a closer look at the film. One is Dr. Hobart van Deusen of the New York Museum of Natural History. Dr. van Deusen, who is president of the prestigious

Explorers Club of New York, made a guarded statement to the effect that it could be real. In conversation with me at his offices in New York he added that as far as he was concerned it was real enough at least to warrant further and serious investigation.

The other man who has taken the time to examine the footage and also to help have it submitted for further examination, is Dr. John Napier, one-time Director of the Primate Biology Program at the Smithsonian Institution, presently Visiting Professor of Primate Biology at Birkbeck College, University of London. Dr. Napier has not only examined the footage, he has written about it and made some very constructive statements about it. As we shall see, they are not all positive. But at least they are statements in written form backed by the weight and authority of Dr. Napier's studies and experience. We shall discuss these to some extent in this chapter. But before we do let us go back to Bluff Creek again for a few moments and have another look at the subject of the '67 footage, its possible size, weight, footprint measurement and stride.

The only measurement that we seem to be able to determine with any degree of accuracy is the length of the Bluff Creek creature's feet. According to Gimlin and Patterson and, later, others who found and measured the footprints on or near the film site and who also made plaster casts of them, the feet were fourteen and one-half inches in length. The stride was also measured at the site and although we cannot be sure of the accuracy of that measurement, it appears to have been close to forty-two inches.

As to height and weight, there has been a lot of wild guessing in this area. Patterson estimated the height at seven feet, four inches. Gimlin believes that the creature was very much smaller than this and probably closer to six feet, one or two inches. As to weight, the estimates vary from 350 pounds-Gimlin-to 800 pounds-Patterson.

In an attempt to accurately determine the weight of the creature of the 1967 footage, two groups of people have examined the site and made series of measurements of the trees that it contains, the tree stumps, logs, etc. Both groups have used human models as a standard of measurement and have shot extra footage of these models, at the site, for comparison purposes. Part of the examination procedure of both groups consisted of superimposing the model used in their footage on the subject of the 1967 footage.

The first group was headed by a Jim McClarin, of Sacramento, California. His conclusions were that the creature measured six feet, eight inches in height.

The second group consisted of myself and one companion as a model. My model-Rick, the son of the aforementioned Al Hodgson, of Willow Creek-was just six feet in height, and using him in superimposition of still photographs taken off the '67 footage and taking very careful measurements at the site itself, I came to the conclusion that the subject of the '67 footage measured seven feet three and a half inches.

The problem-and perhaps the reason for the inconsistencies involved in the measurements taken by the two groups-is that the first group used as their basic measurement guide a frame from the 1967 film that did not show the actual position of the feet of the subject. The second group-my examination-used a different frame that does show the creature's feet.

Napier, in his writings on the phenomenon, gives a great deal of attention to the 1967 footage. He discusses, at some length, the analysis done of the footage by a Dr. Don Grieve, a British anatomist specializing in the human gait. Dr. Grieve wrote a detailed report of his findings after a careful examination of the footage. The figure at which he arrives, for the height of the subject of the footage, is six feet, five inches. Napier accepts the Grieve findings in this respect but points out that if this is the correct height, then there is something very wrong with the size of the figure's feet. The foot measurement given by Patterson, and afterwards confirmed by others who examined the footprints or actually measured the plaster casts that Gimlin and Patterson made of the prints at the site, is fourteen-and-a-half inches.

According to Napier in his book, this measurement is not in correct proportion to the given height. He feels that a figure with fourteen-and-a-half-inch feet should have a minimum height of seven feet, eight inches. He bases this finding on the formula, height = maximum foot length x 6.6. He concludes that either the footprints are fake, or the film is, or both are.

The conjecture and discussion aroused by the 1967 footage is endless. A year from now, perhaps even ten years from now, the subject will be hurled back and forth between the believers and the non-believers. Did Gimlin and Patterson have a partner? Was there a third man with them, a big man who climbed into a fur coat and then paraded up and down while they photographed him? Then who was, is, the third man? Where did he come from? Where is he so successfully hiding now?

Again, did Patterson and associates unknown arrange for this third man without Gimlin's knowledge? Was the third man placed up there in the woods, waiting until he got the signal to move into the creek and then show himself for those few brief moments of frantic surprise that panicked the horses and were over so quickly that Gimlin was bamboozled into belief in a real creature?

There is no doubt in my mind, in the minds of most of the serious Bigfoot "investigators" and, I believe, in Napier's mind, that the 1967 footage is a vital link in the chain of evidence, both hard and soft, that supports the existence of the Bigfoot. Dr. van Deusen, of the New York Museum of Natural History, will agree with me on this. If the footage was faked, it is of course worthless. But if it is real-and we must face the fact that it is possible, it is feasible-then it proves without a doubt that in 1967, in Bluff Creek, Del Norte County, California, at least one female Bigfoot was alive and well and happily striding through the mountain forests.

I was visiting with my brother Bryan in Lake Tahoe in late 1972. We talked about the 1967 film footage and its value as evidence and Bryan pointed out that no one had ever done a really thorough study of the footage with the end in view of definitely establishing the height of the subject of the footage. He suggested that I visit the Bluff Creek area on my return to Oregon, spend some time there, and look for some more answers. I agreed and called Ernie Alameda, one of my associates from the 1960s, who lived in Hoopa, near Bluff Creek, suggesting that he meet me and accompany me to the film site.

Hoopa, where Ernie and his wife Dorothy live, is in the Hoopa Indian Reservation. Ernie owns and manages the Oaks Cafe, in the Hoopa Valley. At least I think that is what his profession is. I am not sure, for I seldom arrived at the Oaks Cafe and actually found Ernie working there. Ernie was a salmon fisherman and invariably he was "gone fishing", somewhere down the Trinity.

Dorothy is a Yurok Native American and is one of the most gracious and charming women that I have ever met. She told me how, many years before, her father had told her about encountered a Bigfoot in one of the little streams that comes down out of the hills to the east of the reservation and flows into the Trinity River. He was gathering crawdads at the time and was down on his knees in the shallow water when the creature came out of the brush, about fifty feet from him, upstream, walked to a big, dry rock in the middle of the stream, squatted down on it with its back to him and defecated into the water. This done it sat for a moment and then slowly turned around and looked downstream, directly at the crouching man.

The point of her father's story, that Dorothy particularly remembered, was that he told her that the creature seemed to sense his presence. He had made no sound and from the moment the thing appeared, he had not moved. Now, making directly eye contact with him, the thing stood up, whirled around, took one big stride off the rock, into the water and with another was gone into the brush.

Ernie was unable to accompany me to Bluff Creek and so, after visiting with he and Dorothy, I loaded my camping gear in my Scout and drove north to Onion Mountain, a peak in the upper Bluff Creek watershed.

There was an old, one-roomed, wooden cabin on top of the mountain in those days and I moved in and made it my base, driving from there, via the Onion Mountain to Lonesome Ridge Road, every day, to the film footage site, to take my measurements and do my studies.

I had with me some tapes of interviews made with Patterson not long after the event of the filming in which he described the incident and the aftermath. These were useful to me in my studies and each evening when I got back to the cabin, after a day down in the creek bed at the film site, I played these on my tape recorder and listened carefully to what Patterson had to say.

From the top of Onion Mountain one can see the Pacific Ocean and the Californian coastline, a splendid sight on a clear day. There is a lot of fog on this coast, even in the summer. During the day it generally stays out to sea but in the evening it moves inland and every day I was there, as the sun sank into the vast Pacific Ocean, I watched it, a solid wave of ivory-colored cream, pouring in across the coastline and through the trees of the lower slopes. At night, when the wind died, there was no sound other than the crackling of my little fire in the cabin wood stove and the occasional lonely call of a night bird.

Footprints were found on Onion Mountain in the late sixties and I could not help thinking, as I lay in my sleeping bag in the silent night, of how many places there were in the vast northwest like Onion Mountain, where a Bigfoot could spend all its life without being discovered by man.

In a week I had finished my studies in the area. I loaded up the faithful Scout and in one long day drove right through to base at The Dalles.

Through personal contact with Patterson and Gimlin, contact that included many hours of face to face interviews and conversations, I was able to apply to my investigation of the 1967 footage a more than average knowledge, what might even be called an intimate knowledge. With my knowledge of the backgrounds of both men

I was also able to ask myself if Gimlin and Patterson were actually capable of creating and perpetrating a hoax-if that is what it was-of the magnitude of the '67 incident. Were they, for instance, intellectually, emotionally, physically and economically capable of the work involved? The construction of the fur suit that is, as seen in the footage, so perfect that it has, after five years, defied the examination of "where is the zipper" seekers all over the world? The costs involved in its construction, whether it was made privately or by a commercial firm. The problems of maintaining secrecy with the firm that made the suit, there being only one or two such companies in the United States. The problems of keeping the third man-the man in the suit-quiet, for the rest of all of his life?

Looking at the footage in this light one might phrase the vital question, how did they hoax it, rather than did they hoax it. In my book both men lacked, primarily, the intellectual capacity essential to the production of a hoax which several examining experts have termed a masterpiece and this, without question, is a point which I believe strongly supports the authenticity of the footage.

There are also other points that help to support the reality of the footage and that came to light as a result of my Onion Mountain research reconnaissance. Two of them that bear serious examination-when the question of a hoax is being considered-are, 1), the physical location of the site in relation to its immediate surroundings and 2), the day on of the week on which the footage was obtained.

The site of the filming (see map) was an open creek bed, approximately one hundred yards in width and with high steep sides, partially covered with timber, partially bare. Just north of the site, the stream bed bends to the left (looking upstream) or northwest. To the south it runs straight for a matter of two or three hundred yards and then curves gently to the southeast. From this curve it continues south-east and then south, down to the junction of Notice Creek and the site of the aforementioned Louse Camp, scene of many a 1960 campfire for me and my companions of those years.

The bed of the stream is generally clear of debris except for scattered logs here and there and two big logjams that lie just to the south of the site of the filming. One of these was the pile around which Gimlin and Patterson rode when they said they encountered the Bigfoot.

In 1967 a road ran along the east side of the stream, close to and level with the water and at places winding in and out of it. It was a dirt and gravel road that had been cut, years previously, by a caterpillar tractor for use in a log removal program. This road was used, at the time, by the occasional fisherman or hunter and it was also negotiable with a four-wheel-drive vehicle.

On the west side of the stream, also running parallel to it but set in its bank about a thousand feet above it, was another road. This too was originally a rough logging road. But in recent years it had been converted into an all-weather road with a bitumen surface. This road was negotiable by vehicles of all kinds and could be driven at moderate speed. Of significance to my description here is that from this road the actual site of the footage could be seen.

This, then, is a description of the actual site of the '67 footage and its surrounding terrain. Open to the south, partially open to the north, visible from the approach road that lay in and along the stream bed and also visible from the road that ran across the hillside a thousand feet above it. It was, generally, speaking, an open area. A highly visible area and, for a hoaxer making a fake film, a highly vulnerable area.

During my research from the lonely little cabin on Onion Mountain, I put myself in the position of a pair of hoaxers intent on making a fake film. Would I choose Bluff Creek? Probably, because Bluff Creek and the forested mountain terrain of upper Del Norte County had, in the 1960s, produced a considerable amount of Bigfoot evidence. The finds dated back to those made by myself and my brother Bryan and Steve Matthes in 1960, and they included at least one sighting and also an incident in which some heavy road-building machinery had been damaged by "something" which, according to at least one of the men who were there at the time, Thomas Sourwine, of Lyle, Washington, picked up a 300-pound stone and hurled it several times against the sides of the machinery in question, In other words, Bluff Creek and its extensive watershed was known as Bigfoot country to many people, a place that had produced credible evidence and where searchers might find more; if they looked.

Again, would I choose the general area between the confluence of Bluff Creek and Notice Creek and from there upstream? Yes, probably, because it was in this particular area, in 1960 and also prior to that time, that most of our footprint finds were found.

But, in having to decide on an actual site for the hoax within this area, would I choose the place where Gimlin and Patterson made their footage? The answer was no, definitely not, and for the reasons that lie in my description of the general area … the open visibility of the area from the south, the north, and the west.

In 1967, when the footage was made, Bluff Creek was still a favorite area for Bigfoot searchers and any weekend one was liable to encounter small groups of people from Hoopa, or from Willow Creek, or even further away, driving in there in their cars, camping overnight, and hoping to get a quick look at a Bigfoot. It was also an area where some deer hunters came and even the odd fisherman. Many of these people coming into the area used the road that ran above the creek on its west side. The road has long since been abandoned (and closed) but in 1967 it was in continuous use. Cars, pickups, jeeps, and other four-wheel-drive vehicles used it all the time and if their vehicles were equipped for it, they could branch off into the bed of Bluff Creek, just above Notice Creek, and take the old streamside logging road that runs in and alongside the bed of Bluff Creek.

A hoax party could too easily be surprised by a car coming up this creek bed road and no one in his right mind would choose the actual site of the '67 footage for time-consuming preparations that would have to precede the actual shooting of the footage, or the actual shooting itself. From the upper road, the one that runs on the west hillside, a thousand feet above the creek, not only could people arriving by car see down into the site of the filming, but a hoax party, working in this area, would probably not be see the onlookers. They would not see them, in fact, unless they were actually watching for them and would certainly not hear their approach above the noise of the water in the stream itself.

Again, there were other potential film sites further up the stream that were much more secluded and where a hoax party would stand much less chance of being surprised. A dozen small streams flow into Bluff Creek in its upper reaches and several of these offered heavier cover, good film sites, and closed surroundings that would give adequate protection for the hoaxers against surprise. One of these, for instance, is Scorpion Creek, a deep, densely forested ravine that I examined during my Onion Mountain research program. It would have made an ideal site for a hoax operation.

As to the actual day, and the time, that the footage was shot, these factors give us two more points that seem .to stand in favor of the footage being real; as such they should be discussed.

 In separate statements, both men said that the time of contact with the subject of their footage was between 1:20 PM and 1:30 PM. Now the actual time is not important. What is important is that it was early afternoon, a time that allowed for the possibility of drivers on the nearby roads, or hikers, or fishermen. In other words, a time that again made a hoaxing effort vulnerable to discovery.

Next, we have the day of the week to consider, in the case of the footage, a Friday, the first day of the weekend, a day when people like to get off from work early and out for the weekly break. It is, once again, the day that campers and hikers, hunters and fishermen and, yes, Bigfoot searchers of the weekend type, pack their rucksacks and fishing and hunting gear and set off for the woods. It would be, for a hoax party, the most dangerous day of the week.

Gimlin and Patterson spent four years doing their part-time searching for a Bigfoot. I somehow think that if the '67 footage was faked, that it must have been something that they planned for a long time and not something that they dreamed up in a few afternoons of riding through the forest. The making of the fur suit alone would have occupied many hours and days of tedious and painstaking work. And surely, when it came to choosing a site for their elaborate and carefully planned fake film, they could have done better than that open creek bed between two roads, where, with the arrival of just one car and the sighting of their activities by just one person, all of their carefully laid plans would have gone down the drain.

And surely they could have picked a better day than a Friday and a better time within that day than the early afternoon? As I stated before, I will not credit either man with the intellect necessary for the making of a near perfect hoax. But I will credit them with intelligence enough to be aware of the unjustifiable risks involved in the use of that place, at that time, for the perpetration of a fabrication.

I am inclined to give the 1967 footage a 95% chance of being genuine and the subject of the footage a 95% chance of being a real living creature. I base this belief on my own examination of the evidence that the footage supplies, of the footage itself, which I have viewed probably a hundred times, and of hours of personal research at the site. I am inclined to disregard statements such as "nothing walks like that, therefore it is a fake" in spite of their supposedly learned scientific origin. I am equally inclined to refute Dr. Napier's statement (much as I respect his qualifications), about the size of the feet being totally out of proportion to the size of the creature. I believe the very basis of this statement, Napier's formula, height of figure equals length of foot times 6.6, to be totally inappropriate to the particular subject. How can one apply a known formula to an unknown quantity? Why should a totally unknown, unclassified, never examined creature of six feet, five inches not have fourteen-and-a-half-inch feet? Or ten-inch feet? Or twenty-inch feet? Where is the rule by which we measure the unknown in this case? To my mind there is none.

Applying this formula is exactly the same as saying that because it has hairy breasts, and apes do not have breasts, and humans do not have hairy breasts, it must have been a fake. Why should it not have hairy breasts, if it is a creature that is totally new and completely unknown to science?

I would be a lot happier about scientific opinion on the '67 footage if it were based on thorough examination of the footage by accredited experts under controlled laboratory conditions. But it is not and much as I respect the opinion of men like Dr. Napier, I feel strongly that, where the subject of the 1967 footage is concerned, scientists in this country should be talking less about fakery and more toward solutions of the enormous questions that this unique strip of film pose.

Conclusions, Superstitions, Scientific Support and the Probability of Existence.

Native American people of the USA do have a number of beliefs in superstition. There may be some attached to the belief in the Sasquatch. But It is limited and in all of my research, which has included considerable contact with Native Americans, I have encountered very little and what there was I saw as widely separated from the Bigfoot mystery. The creatures they call Bigfoot, or Sasquatch-and many other names-refer to a real creature and when translated into English do not emerge as giant frogs, or fifty-foot snakes, or monstrous birds that can breathe fire, but as real living creatures, manlike beings, walking upright and covered with hair.

One of the Native American people who worked with us in the sixties, the aforementioned Betty Allen, talking with me about the reality of the Sasquatch and how separate it was from the superstitions of her own people, the Hoopa tribe, gave me a sample of one of the of the beliefs of her people. I recorded what she said to me in my journal for August 15th, 1960 and it reads as follows:

There has always been an Indian devil living in Bluff Creek. But in Mill Creek there were wild men and every hunter going in there to hunt had to add a twig to a big pile near the main road to safeguard his hunting and his return. Native American Sasquatch researcher Betty Allen, Willow

Creek, CA, to the author, August 15th, 1960, as recorded in his journal for that year.

In summary of the claims of some-including many scientists-that the whole Sasquatch odyssey is nothing but myth, I would beg to differ by stating that I see the gap between myth and reality as a big one and very obvious.

And so we come to the last chapter of the EPILOGUE, and, as its title says something about conclusions, it looks as though the author is going to have to make a few. He will try and wriggle out of it of course. He will beat around the bush, make excuses and generally try to confuse the reader with literary parallelograms and other deviousness. But in the end the title of the chapter will pin him down and he will be forced to put in writing, for all to see and for the skeptics to play merry hell with-for a time at least-his own conclusions on the reality of the Bigfeet and on what should be done about them.

The skeptics apart, he is going to have to do this out of respect for his readers, many of whom have no doubt paid good money for his book. (If they have got as far as this last chapter then they must have bought the thing, he thinks; so let's avoid irritating them any further by at least bringing it to a decent ending.)

So, let us summarize and conclude and start by looking back at the four areas of evidence that seemingly support the existence of the ubiquitous creature of our interest, These are, the history, the footprints, the sightings and the 1967 Bigfoot footage.

To start with the history, I think that we might be stretching a point if we allow the Norsemen's encounter with creatures they described as "horribly ugly, hairy, swarthy and with great black eyes" to be put forward as evidence. But Leif Erikson's thousand year old story has a colorful ring to it and so we shall leave it with the ages and comment on it no further.

The remainder of what we see as a historical background, e.g., the accounts of the various incidents that are recorded through the 1800s-the Yale, British Columbia capture of what might well have been a young Bigfoot, the 1924 Ape Canyon incident, and the many other stories that I and my associates have managed to unearth in our research work, all seem to support the actuality of the existence of the creatures.

Where authentication of these old stories is concerned, however, there is a of course a problem. The people concerned with them have long since passed on and in many cases there is only the briefest of records. With the Yale, British Columbia, story for instance, as an example, there is just the single newspaper article. After that there is nothing, and to this day no one knows what became of "poor Jacko," whether he was shipped to England, as the owner talked of doing, or whether he succumbed in his cage to the no doubt unmerciful attentions of a gang of tough, north woods railway men. Thus, all that we can say of these early stories is that most of them and certainly the ones that have been described in detail here-do seem to genuinely refer to the creatures that we know in the present day as Bigfoot.

The second area of evidence is the footprints. Before this examination, let us take a look for a moment at what is meant by the word "footprints," where we refer to it as evidence. It means, in this context, large, five-toed, humanoid prints of unknown origin, usually found on mountain roads in the coast ranges, in isolated areas, on surfaces that include sand, mud, dust, shingle, snow, or grass. The prints usually contain a heavy impression suggesting great weight and they are separated by a stride that is longer than a normal human stride.

Credible prints that have been found seem to average from fourteen to sixteen inches. I have heard of larger prints, but I have never seen one. I have been told of twenty-four-inch prints being found but personally do not believe in prints of that size being real. It is my opinion that twenty-four-inch supposed Bigfoot prints that have been reported in snow are the result of melting. Footprints in snow will melt out from two inches (e.g., deer hoof imprints) to ten inches or more within a few days. Again, odd conditions of air movement and temperature may cause the opposite effect and ten-inch prints will sometimes shrink to four inches or less. It is important to remember, however, that there is never continuous uniformity of size and shape to prints that are the result of melting.

The average stride length (heel to heel) with fourteen to sixteen in prints is from forty to forth-five inches. For the interest of the reader, the author is six feet in height and has a normal stride of thirty-three inches.

The prints, when found, seem to indicate a huge weight, and although it is nearly impossible to determine this weight, using a print, with any degree of certainty some of the prints that I have personally seen, impacted in sand and mud, had a great weight behind them, a weight that I believe was in excess of 400 pounds.

There is no doubt that many of the footprints that people find in the north west forests and that are taken to be Bigfoot prints are made by bear. Because of this, I think that we should have a look at the bears for a moment. There are a lot of bears in the Pacific Northwest, thousands in fact, and every time a bear moves more than a yard or two he lays down at least four paw prints. Euarctos, the common American Black bear, comes in different colors and different sizes. Some are brown, some are cinnamon, some are black. Regardless of color, however, the one bear common to the United States and parts of southern Canada is the Black Bear.

There is no other species of bear as numerous, as prolific and, with his different colors, as confusing to the public as the Black bear. When cinnamon he is often confused with the grizzly and when brown with the Brown bear. The Black stands about two to three feet at the shoulder, weighs in at 200 to 450 pounds and has rear paws-the ones with which we are concerned here-that seldom exceed six inches in length. The grizzly, on the other hand, is a very much larger bear, stands three-and-a-half feet at the shoulder and weighs up to 1100 pounds. The Brown, the largest bear in the world and the largest meat-eating animal on the face of the earth, can be four to four-and-a-half feet in height and can weigh up to 1600 pounds.)

The front feet and the rear feet of all bears are quite different in shape. The front is shorter, has a rounded pad, and its imprint is a pear-shaped indentation crowned with the circular toe prints of five toes. The forefoot imprint very often, though not always, may also has claw marks.

Its rear foot is almost twice as long and is very human in shape. There is a narrow heel and an instep, and when the print is capped off with five toe marks, the result is often mistaken for a human footprint. And, when the person who finds it has the Bigfoot "bug" and is far out in the mountains, with the wind moaning in the pines and the nearest human habitation far away and a Bigfoot lurking behind every other tree, the print is invariably taken to be that of a Bigfoot.

Reports of sightings and footprints come in regularly to the Information Center in The Dalles. About 75% of the prints that are found and that are reported to the Center as genuine Bigfoot footprints are Black bear paw prints. They are usually found by people who honestly believe them to be Bigfoot prints and who call the Information Center in good faith. People calling or writing the Information Center with information on the phenomenon are always thanked, and as far as is possible the information is always acted on by the Center. For the public is the main source of information from which the Center draws its investigative material. An experienced woodsman would not confuse a bear paw print for a Bigfoot print, or, for that matter, with a human footprint. But the average person may well do so, and in the excitement of a find the inexperienced are apt to forget that 95% of all the Bigfoot prints ever found have been more than ten inches in length, while the Black Bear seldom has a rear foot that exceeds six inches.

Thus, while the Black Bear does make a Bigfoot-like print, I do not think that the American species can be blamed for what are normally believed to be real Bigfoot prints. (At the same time I do think that they can be blamed for making what have been mistakenly called Bigfoot beds.

The female Black, when about to have cubs, often makes a most elaborate "nest.'" This will be roughly circular in shape and up to six feet in diameter and will be constructed of carefully laid-almost woven-sticks, moss, leaves and grass. The result looks for all the world like a huge bird's nest and several of these been found by Bigfoot hunters who believed them to be Bigfoot beds, or lairs.

The grizzly bear has a hind foot that is similar in shape to that of the Black bear of the United States, but of course his paw print is very much larger. Could the grizzly be blamed for some of the huge humanoid footprints that have been found in the United States and that have been believed to be Bigfoot prints? Yes, if there were grizzlies in the United States and in the places where the prints have been found. There are still grizzly in some parts of the U.S., in Wyoming, possibly in northern Idaho, and just possibly, occasionally, in northern Washington. But there are none, any more, in Oregon or in northern California.

The grizzly, a big, short-tempered, easily provoked bruin, is, like the American Indian, a creature that barely survived the white invasions of the nineteenth century. Hunted and shot, harried out of existence, he disappeared from nine-tenths of his former range and is today, in this country, an endangered species. Again, the grizzly has a set of thick and powerful nails set in the toes of each foot. It is possible for the animal to walk without showing any nail impression, for the nails are raised above the level of the foot pads. But this is only when its walking surface is very hard soil or rock. In mud, sand, snow, or heavy dust, the claw mark will always show and no claw marks have ever been found in what have been regarded as genuine Bigfoot footprints. The grizzly, like the Black, is thus ruled out as the villain behind the scenes where Bigfoot footprints are concerned. Which leaves us with the question, who or what makes the footprints?

There is no doubt in my mind that some of the footprints that I have seen, which others regarded as real Bigfoot footprints, were faked. This includes some of the footprints that we found in California in 1960. The falsity of some of those latter findings was confirmed by Steve Matthes-an expert with tracks-and expert is a word that I use with care, who declared them fake and showed how and why he thought so. There is equally little doubt that some of the footprints that others, less expert, have found, particularly in domesticated areas, have also been faked, some of them made with a care and attention to detail that convinced many people that they were real, others created with crude wooden feet, or with thick, rubber-soled shoes, the soles of which had been cut to the shape of large human feet.

At a recent meeting with the Anthropological Department of Portland University, where I spoke on the subject of the phenomenon, a medical doctor, supposedly a friend of mine for a while, produced a pair of stiff-soled, crudely carved wooden feet and proffered these as evidence to prove that all Bigfoot prints were faked! Prints, made from the set that he so proudly popped out of his little medicine bag, would have been quite ludicrous in appearance and good for nothing but a laugh or two.

One thing that adds to or detracts from the credibility of a footprint find is its physical location. It is an established fact that the creatures of our interest stay well away from human habitation. Thus, footprint "finds" in towns or in and around dwelling houses, instantly lose much if not all of their credibility and more often than not can be quickly classified as fabrications.

On the other hand, finds in remote areas always have much more substance to them, if only in consideration of the fact that a faker is simply not going to hike back into an area and make prints where no one may ever find them. Such was the case with a set of prints found by myself and my brother in the winter of 1960.

One day in the winter of 1960 we drove the Scout into the Hoopa Indian Reservation and after a stop for coffee at the Oaks Café, and a chat with our friend, Ernie Alameda, the owner, we left and drove due east, up the high ridge that lies to the east of the reservation.

We used an old logging road and when this came to an end, left it and drove on hard snow up an access road to a big clear cut ion the forest. The snow was hard but the crust kept breaking and only the Scout's four-wheel-drive and low-ratio gears got us through. At the end of this second road we left the Scout and walked on up the ridge, a distance of about one mile to find, on the top of the ridge, a set of fifteen-inch, five-toed, humanoid footprints, fresh and imprinted half an inch into the snow crust.

We examined the prints very carefully and checked the snow in the vicinity for boot prints or other sign of human presence, finding none. The prints looked very real too us and their location, on top of this isolated ridge, some twenty-six miles back in the mountains, added to their credibility. For not only had we told no one where we were going that day we never did, if only to try and keep one jump ahead of hoaxers-but when we set out that morning we had not had any particular destination in mind. In fact, we had arrived on that ridge merely by chance and it was then, as it is now, extremely doubtful to me that a hoaxer would have made prints out there in the winter cold, so far from human habitation, in an area so isolated, just in the hope that someone would find them. And so we judged the footprints to be genuine and to have been made by a Bigfoot.

Today, many years later, I feel much the same about many of the footprints that have been found in the mountains in the last ten years. A few have been faked, but for others there is no other explanation than that they were made by a long-striding, big-footed biped of great weight, something with a humanoid foot that was not a m an.

Next, the sightings. How many of them are real? How many faked? How many of them imagination? How many of them are concocted encounters created to boost the sorry egos of fakers and hoaxers.

Let us first examine the probability of fakery in this area. Briefly, there has been "sighting fakery" where films are concerned. For instance, it has been quite conclusively proved that one infamous piece of film, shot in the vicinity of Evans, Washington, in 1970, was faked. A second piece, examined at Yerkes Primate Institute, Emory University, Atlanta, made by the same person, was also faked. (This latter film showed a stumbling white figure in a baggy fur suit that at one time, I was told, waved at the cameraman.)

This means that there are, (at this time) , to use the term, "gorilla suits" being made and being used by at least two hoaxers. The suits have been used in films and will in all probability be used again. But to use a fur suit in a fake film where one is being photographed by one's wife, or one's partner, under reasonably secure circumstances and surroundings, is one thing. To put on the same suit and go capering up and down some lonely road in the middle of the night, or across someone's meadow or farmland, or up on a high, open ridge, is another kettle of fish altogether, because the Pacific Northwest is gun country, a place where, quite legally, people carry guns a great deal of the time.

It is not unusual to see people carrying weapons, both pistols and rifles, in their cars when they are out on both public roads and back roads. The "easy rider" type pickup, with the rear-window gun rack, is a common sight in the back country of the northwest and when the rifles are not visible on the gun rack, they are usually behind the seat or underneath it. If there is no rifle there is generally a pistol lurking somewhere, either on the seat, or under the seat, or in the dashboard compartment. The given reason for all of this armament is usually protection. But in actual fact it is something else. It is a carryover of the great Western gun cult that sprang to life in the wild days of the first miners and mountain men and that today has developed into a tradition that is an accepted part of life in the northwest.

When I first came to the Pacific Northwest, to commence my investigation of the Bigfoot phenomenon, many people asked me if I intended to shoot one. I said no. That was, and still is, my intention, and part of my work here and the work of my associates has been to strenuously oppose the "shoot it to prove it" gunmen who would kill one to serve their own ends. At the same time there is undoubtedly a protective feeling among many people in the northwest concerning the Bigfeet. We have encountered it again and again in the concern of people that harm or injury might come to one of the creatures, creatures that in their knowledge have never harmed anyone. At the same time there is now, 1975, an unfortunate movement among a few of the gun-toting community to have one shot, if this is possible. The movement is propagated by the vague promise of some huge reward from some equally vague institution and among some people this ethereal promise has now become fact. There are no details of the amount of the reward and the institution that will buy a Bigfoot carcass and then happily hand over a million dollars or so is not named. It is not named because it does not exist. But this promise of huge monetary return, strengthened by a recent newspaper article about a Texas millionaire who would pay half a million dollars for the body of one of the creatures-there is no such person-has in some uneducated quarters grown to the point where it is firmly believed.

The result is several gun-carrying Bigfoot hunters who have openly stated their intention of shooting one should they get the chance-their lame excuse being that it would not be fair to expect them to turn down the chance of a million dollars.

The hoaxers that we have encountered in our Bigfoot research have been people who were generally aware of what activity was taking place in the Bigfoot field. They were aware of the stated intention of the few gunmen who hunt the Bigfoot with rifles and they were equally aware of the number of guns carried by people in the northwest, the ability of those people to use them and the risks involved in walking around in the woods in a fur "gorilla" suit.

They also knew that the hoaxer who is going to do this, pretending that he is a Bigfoot, is not only asking to be shot but very probably is going to be shot, particularly if he persists.

Any man who does this and who is aware of the risks involved is either a fool or a madman. There are, no doubt, hoaxers living in the northwest or in Canada today who will do almost anything to satisfy their peculiar cravings for notice. But I know of none who will knowingly and continuously face the probably fatal risks involved in "gorilla suit" fakery at this level.

Next, how many of the sightings are imagination? How many people see things on lonely roads at night, in the lights of a car, or far ahead where their car lights barely reach, that they think are Bigfoot? Tree stumps, clumps of bushes, telephone poles? The answer, quite a few.

One man called us to say that the night before he had very definitely seen a Bigfoot, standing with its back to him, streaked grey and black, huge, bent over, broad-shouldered, head down, unmoving. He had stopped his car, backed up, and it was gone. Investigation at the place of his "encounter"-he came with us to show us the place-revealed a big roadside stump. The man, who honestly thought that he had seen a Bigfoot, was quite astounded when we showed him the stump and his tire marks, where he backed, almost, but not quite, up to its base.

One woman called us late one night. Her son had seen a Bigfoot crossing a road near The Dalles. We investigated and found small bear paw prints arriving at one side of the road, crossing the road and then leaving from the other side, at the precise place where the boy said that he had seen a "large hairy figure." When we showed him the bear tracks he refused to accept them, suggesting that we were wrong and that they were Bigfoot prints.

Imagination in an individual is, I believe, in direct proportion to the individual's measure of intelligence, level-headedness, and plain common sense. Some of the sightings reported to us have been the result of total imagination. But not all of them, and certainly not, to my way of thinking, not all of those that I have described in the preceding chapters.

How many of the sightings are simply fabricated stories, concocted by the ego-hungry individual as a means of supplying something missing in his life? Answer, not a few. But the fabricated story is usually fairly easy to expose especially if, like the members of our Dalles Information Center team, one has a background of experience and knowledge that can be used in the questioning of a "suspect." A knowledge of the terrain, for example, is very useful when questioning someone about a particular area where they may say they had a sighting.

One man, for instance, told us that he had once seen a Bigfoot about half a mile down a canyon in the central Cascades. He was riding a horse at the time, he said. The expression, "down the canyon," struck us as odd in his story and I asked him which way he had entered the canyon. He told us, after a slight hesitation, that he had entered the canyon from its upper end and that he was riding downstream when he saw the Bigfoot. Unfortunately for the credibility of his story, we had been in that canyon only a few months previously and we knew that its upper end was enclosed by a sheer rock wall several hundred feet high.

Another man told us that his sighting took place high on a mountain ridge and the Bigfoot that he saw was moving at a distance of 350 to 400 yards. It was never closer than 350 yards and he saw it without binoculars or a rifle scope. But he was able to tell us the color of the creature's eyes (dark blue, he thought) and also to notice that its fingernails were thick and broad and flat. Fingernails, like eyes, could not be distinguished for color or shape by the human eye, unassisted by binocular magnification, at a distance of 350 yards. A story of a sighting at this distance, that included these details, could only be construed as being a fabrication. This kind of story, like the canyon story, simply goes into the storage files after which we waste no more time on it.

I think that most people will agree that there are not a few fabricated sighting, stories that obviously have no basis in truth. At the same time I do not believe that this detracts to any great extent from the accounts given of sightings by people of reliability and integrity, men like Bill Taylor and the Welch brothers of British Columbia and others of their standing and morality.

This brings us, in conclusion of our examination of the Bigfoot supporting evidence of the reality of the phenomenon, to the 1967 footage. We have already discussed this to some extent, so let us recap the pros and the cons on the filming and the footage.

The skeptics, who say that the 1967 footage is a hoax, put forward a very unconvincing argument. The basis of this argument is that the Bigfoot does not exist and that therefore the footage must have been hoaxed.

Among the skeptics, as I have mentioned previously in these pages, are many scientists, and their reasoning why the footage must be faked is summed up in the only two statements that I have been able to personally hear from them on the subject.

These are (1) "They do not exist because if they did exist we would have known about them by now." Personally I think that it would be very interesting to apply this statement to the Tasaday, of the Philippines. "The Tasaday do not exist because if they did we would have known about them by now." Or to the coelacanth, the fossil fish that was discovered, alive and well, off the coast of east Africa, just a few years ago. "The coelacanth does not exist because if it did we would have known about it by now."

(2) "Nothing walks like that and therefore something walking like that must be faked." A fascinating statement and all that I can say about it is that I am glad that I do not have the mentality that went into the

creation of that utterance. For surely even a halfwit can see that if the Bigfeet do exist, and if they are indeed a totally new and unknown species of giant primate, that they obviously do walk quite differently from anything known to science.

Surely any scientist can see that if the Bigfeet do exist, they are entitled to have a neuromuscular system totally different from the human system. That being different, they are not obliged to walk in the same manner as any other primate, including us, regardless of what the rulings of anthropology and biology.

To me it does not matter if the subject of the 1967 footage does not straighten its leg in stride, as the footage seems to suggest. It does not matter if it has a forty-two-inch stride which is scientifically believed to be wrong for its height. It does not matter if its feet are too big or too small, in proportion to the size of its body. It is obvious, if the thing in the 1967 footage is real, that it is something completely different from anything ever seen before, ever known to science. Something that is totally new and unknown and thus, by its very newness, is entitled to look different, entitled to move differently, entitled to be different in all of its physical behavior, from any other animal on the face of the earth.

The 1967 footage is further discussed in the appendix of this book. For myself, I am inclined to give it a 95% chance of being real. I allow it this because of the reasons which I have already discussed. They are not scientific reasons. They are a layman's reasons and they are based on personal contact with the people who made the footage and my knowledge of how they made it, of their intellectual capabilities and of the circumstances surrounding the production of the footage.

If the 1967 footage is a fake, then it is a masterpiece, and I think that I can sum up on my belief in the validity of the footage by saying that it is mainly based on my inability to understand how the two men concerned could have created that masterpiece. I give it 95% credibility and I reserve 5% to apply to the possibility that I could be wrong. But it would not take too much to have me drop that 5%.

Toward the end of his book about the Bigfoot and Yeti phenomena, Dr. Napier, having gathered all of the evidence and analyzed it in an intelligent and scientific fashion, arrives at the point where he feels that he must come to some conclusions and, as a writer, put them down on paper. He does this in an ambiguous statement which, I believe, leaves his readers very much up in the air as to what he really means. He writes that he is convinced that the creatures exist but thinks that they are "not all that they are cracked up to be." I find this statement confusing. For me, with the exception of his 1967 film examination and his views on the lack of food in the northwest for a primate as large as the Bigfoot-it clouds a reasonable analysis of the phenomenon. A statement of conclusion should be more definite than this.

The evidence that has been examined in these chapters is not all of the evidence extant in support of the Bigfeet. There is a great deal more, in fact. What has been presented in these pages is the more solid evidence that I and my team of associates have been able to uncover and examine and analyze in our five years of investigation. It is not, and I shall be the first to admit this, what the scientists call hard evidence. But nevertheless it is evidence and I feel that all of it together presents us with supportive material that is well worth the time and money that was spent on its finding. If nothing else it has allowed us-I and my associates-to make some very definite conclusions on the phenomena as a whole.

How have we gone about collecting this evidence? In all the ways I have discussed:, which is examining footprints, investigating reports of sightings and dedicated full time research, one example of which is the surveillance we conducted of a potential area in 1972 and 1973. This was done with the assistance of the Academy of Applied Science of Boston, and Dr. Robert Rines. An account of the surveillance project and how we conducted it seems worthwhile here.

It was through the good offices of my friend Tim Dinsdale that I first met Robert Rines and, through him, the fascinating group of people who were associated with or actually taking part in the Loch Ness project. Tim took a lively interest in the Bigfoot phenomenon, and urged me to fly to Boston to see if the Academy of Applied Science-his own part-sponsors- could become involved in my search.

Bob Rines, who was later to become a personal friend of mine, is a Boston patent attorney. He is also Dean of the Franklin Pierce Law School in Concord, New Hampshire, and President of the Academy of Applied Science, chartered in March, 1963, as a nonprofit, scientific and educational corporation, under the laws of the Commonwealth of Massachusetts. The A.A.S, as it is known, is an institution that is concerned with fostering cooperation among the creative individual, industry, education, law, and government. What is called the historical purpose of the Academy was outlined in one of its first monographs, published in 1972. It reads:

To bridge the gap in communications, understanding, and cooperative effort (particularly to attach problems requiring simultaneous interdisciplinary inputs and approaches) between the engineering, technological, innovative and applied scientific community on one hand and industry, university and legal and government institutions, on the other.

Since 1963 the Academy has been engaged in programs in very many areas within the United States and also in Taiwan, New Zealand, and westward through Asia, the Middle East, Europe, and the United Kingdom. The activities of members of the Academy throughout the world have been widely diversified. One of these has been the investigation of the Loch Ness Monster mystery.

In Boston, on that first visit, I was honored by being asked to speak to members of the Academy at the Academy offices in Belmont, a suburb of Boston, and further honored, later, by being made a member of the Academy. Robert Rines introduced me and I talked for about thirty minutes on the subject of the Bigfoot. The talk was well received and afterwards Bob told me that he believed that he could persuade the

Academy directors to start taking an interest in my project and give it some support.

The result of this was a visit to The Dalles, soon afterwards, by he and Tim Dinsdale. They came to make a personal evaluation of the phenomenon and to see what action I was taking in my search and investigation and how they could assist me with planning and with equipment.

They were particularly interested in The Dalles, and together with Dennis Jenson, we walked out most of the area where the sightings had taken place over the years and where we had our observation post. They stayed in The Dalles with me for several days, working on plans for upcoming summer surveillance from the observation post.

Tim thought that we might try to get nearer to the hill of Crate's Point, the hill down which we believed the Bigfoot came in his periodic visits to Hidden Valley. He suggested a route up the back of the cliff above what is called Tuli Terraces, a trailer court area that lies north of the hill itself and skirts the old Dalles to Hood River Highway. We thoroughly discussed the problems of concealment, and with Tim's experience at Loch Ness and my own experience with the problems of getting close to large dangerous animals, we worked out some new plans for the observation of the Hidden Valley meadows and the hill face of Crate's Point. Bob Rines, who as well as being an attorney is also an MIT physics graduate, explored the problems of actual surveillance. I had hoped, for this season, to have enough people working with me, volunteers and local helpers, to keep a watch on the area by both day and night. Diurnal surveillance posed no problems, but night watching was another matter. With powerful binoculars-and we had those-we could see at night over the five to six days of the full moon, at distances up to 500 yards. Without the moon, however, this visibility was reduced to one hundred yards or even less, depending on clear skies and starlight. Bob thought that the answer could lie in electronic night vision equipment, and he promised to give some thought to obtaining some, via the Academy and possibly from the U. S. Army.

To round off their visit to The Dalles, Dennis and I took our visitors out for a short camping trip. A study of our charts showed that one area, in the mountains south of the Clackamas River, had a potential for sightings at this time. So we loaded up the Scout and drove there and made camp.

It was still fairly early in the year, with low night temperatures and we so had to lend our guests some long-john underwear and insulated jackets. When the sun went down the high mountain forest quickly became cold and dark and I remember on our first night seeing Dinsdale, unaccustomed to camping in such rugged country, looking around him nervously, out to the gloom beyond the campfire where, for the uninitiated, there is often a Bigfoot behind every tree. Later he admitted to being a trifle uneasy and I recall that we found this amusing. But I wondered how we would have felt taking part in some of his lonely watches on the Ness, floating on that huge, silent lake in that grey Scottish fog, the nearest help several miles away, and underneath, separated from one's feet by only half an inch of fiberglass, a thousand feet of cold, black water. For us, I feel, there might well have been a lurking monster behind every fog wraith, a huge serpentine beast just waiting to come boiling up out of the depths, to seize us and drag us down into its watery lair.

Bob Rines returned to Boston and Tim went back to England to continue his search. In Boston, Bob started making inquiries into the possibility of obtaining some night vision equipment, and early in 1973 he found what he was looking for and called me. He had heard that I was going to be in New York and wondered if I could make a quick trip to Boson to see him, before returning to Oregon.

I had been honored by an invitation from the Explorers Club to be a guest speaker at their annual dinner at the Waldorf Astoria. My subject, of course, was the hunt for the Bigfoot and for the occasion I had put together a short 16mm movie that included the footage from Bluff Creek, California, and scenes from the search activity in The Dalles and elsewhere. The Explorers dinner was a grand affair and I gave my talk before an audience of one thousand members from all over the world.

A few days later I flew to Boston for talks with Bob Rines. There, at the Academy offices, he told me that he had been able to obtain the loan of some very high-powered night vision equipment from the U.S. army and that it would be ours for a full month. In addition, there would be some medium-powered viewers, smaller apparatus that could be used manually. All of this would be available through the May-June period that we considered crucial in The Dalles area, for the full period.

This was great news. He added that two men would be accompanying the equipment. One would be a technician from the Academy whose job it would be to provide maintenance and also, should we wish it, to take part in the watch. The other was an Army sergeant, whose duty it would be to guard the equipment against theft or loss.

The equipment duly arrived and with it the two men. The technician was one Warren Robinson, a genial fifty-year-old whose knowledge of night vision equipment of all kinds was the result of years of work and experience. Mr. Robinson quickly became "'Robbie," and on his arrival he indicated his willingness to take a full part in the watch from the observation post and in any other work in connection with the project. His military companion was a Sgt. John W. Cannady, a night vision expert, a southerner and a career soldier of serious mien.

The equipment consisted of two hand-held Starlight Scopes, each with a range of about 500 yards, depending on light sources. For short-range work there was a hand-held Thermal Device. And for installation in the observation post there was a massive Night Observation Device, with a 40mm tube and, as we discovered, almost unlimited range.

We started out our surveillance, of what we called our area of potential a few days after Robbie and John arrived. Our plans were to include them in the watch from the hill and so, without delay, we installed them in a comfortable motel in the city and briefed them on their duties.

As their actual working equipment was for night use, I planned to have them accompany us only on the night watch. This meant heading out at six in the evening-1800 hours to our Army men-and staying in the observation post until nine the following morning. During the day they could rest and have the daylight hours free. This plan worked well. With the help of a volunteer group that included my part-time assistant, Lyle O'Connor of The Dalles, Jim Day and Darrell Buckles of The Dalles and Nick Bielemeier of Hood River, we managed to break the night watch down to actual work periods of no more than three hours maximum. Daylight watches were somewhat longer. We found it easier to stay awake during the day.

The night vision device that Bob Rines obtained for us was called a NOD (Night Observation Device) 40MM and it consisted of a single huge scanning lens, about 15 inches in diameter and a heavy, three legged, metal tripod.

The observation post we built on the top of Table Mountain was constructed of heavy wooden 4X4 beams and local rock, with a canvas roof to protect against rain and a sod covering on top of that for camouflage. It was ten feet long, six feet deep and, to allow for two men to be comfortably seated inside, four feet in height. Its front opening was a designed to give us a full view of the area in front and below us and the NOD was set in the middle of this opening with a 160 degree angle of view.

Previous to building the observation post we had drawn up a chart of all of the area of surveillance and from this knew the distances to all of the principal landmarks.

On the first night in the post with the new apparatus, Robbie demonstrated to us the extraordinary capability of the big NOD. There was no moon, only starlight, and this reduced the limits of ordinary vision down to about a hundred yards. For his demonstration he chose a small poplar tree directly in front of and below the post, standing at a distance of over a thousand yards. Through the NOD, with nothing but starlight for power, we were able to see this tree quite clearly and later to actually photograph it through the device. We found that we could also see all the way up the face of Crate's Point Hill, distances of from two to three thousand yards and even beyond that to the highest point of the hill, the 2020-foot ridge, at a distance of five thousand yards. True, at that distance a man, or a manlike figure, could not be seen on the screen. But larger objects would show clearly and in the meantime we were satisfied to be able to visually pierce the darkness all the way across our surveillance area.

One night we watched a large bull elk come down the face of Crate's Point Hill. It stayed out in the open at the north end of Hidden Valley, grazing, and later, in the cold hours before the dawn, climbed up the hill and disappeared into the ridge's oak forest. Another time a group of seven deer came out into the meadows below the post and commenced browsing. We watched them on the NOD and then switched to its secondary devise, a Thermal scanner. On this they showed up as small green mushrooms, their pattern slowly changing shape as they moved around.

Late one night, as we kept our silent vigil from the hilltop, a car drove into a private stone quarry, below us at the base of the hill. The vehicle moved slowly through the quarry and then parked, lights off. The quarry owners, who lived nearby, saw the headlights as the car came in and, suspecting a prowler, called the Sheriff's Office. The car left before the Sheriff's patrol could arrive and we watched it on the scanner as it drove off in the direction of Hood River. Next day, through the local grapevine, we learned the driver's identity-a gun-toting Bigfoot hunter from Canada-one of the mindless clique who openly stated their intention of getting a Bigfoot by the simple method of shooting one on sight.

Next night, to our surprise, he was back. Again it was very late. We watched as the car slid quietly into the darkened quarry and as the driver switched off his lights and parked, we aimed both the NOD scanner and the Thermal Device at him. We were not too worried about his disturbing our area of potential, one which at that time of night was otherwise dark and very quiet. What perturbed us was the high-powered rifle that he was known to carry should he see "man-like" figures on the hilltop-us! And the possibility of his shooting in our direction. We watched him for about five minutes. The Thermal Device registered, on its flickering green screen, the car's engine and lights and also a curious point of heat that seemed to lie within the driving compartment. Then the Sheriff's Department patrol car, driven by Deputy Sheriff Bob Hazelett, came racing into the quarry and pulled up behind the prowler. The deputy stepped out and walked forward to the parked car and ordered the driver to step out. His license was examined. He was advised that he was trespassing and told to leave at once, which he promptly did. Later I asked Hazelett what the other heat spot within the car could have been. Our gun-toting prowler, he told us, had been smoking a pipe.

In the months of May and June, 1973, to his old stamping grounds in The Dalles, the Bigfoot came not. Why? We do not know. We know why he did not come in 1972. In that year, during the month when the sightings had taken place, another group of Bigfoot hunters' from Eugene, Oregon, all armed to the teeth and with at least one large German Shepherd roaming loose around their bivouac, made a camp right up on the Crate's Point ridge, the most likely Bigfoot route into area. Each night they burned a huge fire and generally made so much disturbance that no Bigfoot in his right senses would have come within a mile of them. During the day the dog ran free and two of campers wore brightly colored shirts which on one occasion were detected by myself and my assistant, Lyle O'Connor, without binoculars, at a distance of 3000 yards! Fortunately for all, after about ten days they left.

Through '73 we consciously continued our surveillance. We carried out night watches and day watches and tenaciously peered through our instruments from our Table Mountain observation post. The weather in late May and June was cold at night. Then it turned hot. Then cold again. Our real weather enemy however was the wind. The lookout post faced north west toward the mouth of the Columbia gorge down which, every night after the sun was gone, the wind roared without ceasing; at times the post shuddered on its very foundations.

The instruments, particularly the big scanner, had to be firmly tied down to protect against the constant gusts that hit the structure. When it rained, which it did from time to time, the drops came like bullets through the front opening. Constant repairs were necessary as the nightly wind tore at covers and fittings. One night a gust took the back door off its hinges and blew it five hundred yards, down into the quarry below us.

We had one violent storm during the surveillance period. That night I was on the hill with young Mike Kuhn, our youngest volunteer. We watched huge black clouds racing up the Columbia River gorge and listened to the wind beginning to howl. Then vivid lightning began to flash across the gorge from the Washington side, and when the first bolts began reaching long blue fingers towards our site, we decided that discretion might be the better part of valor and made a hasty departure for safety. We were back with the dawn.

FOOTNOTE: In June, 1974, a violent storm, with 80 MPH gusts, completely destroyed the post, blowing the stone roof off and into the quarry below. Deemed unrepairable, it was then abandoned.

During the watch we made frequent reports to Bob Rines in Boston, kept a log and also continued our investigate work. Then the surveillance came to an end. Robbie and John returned to Fort Belvoir. Our volunteers dispersed. Lyle O'Connor left to join the Air Force, and about the same time Jim Day flew to San Diego and joined the Navy. It was a disappointment to all of us that our determined and prolonged effort had produced no results, just as it was to Bob Rines and the members of his Academy. After the considerable effort they had made on our behalf and the support that they had given us it was disappointing to all concerned that there were no new findings.

At the present time, the Academy is taking an increasingly active interest in the search for the Bigfoot. Their activities at Loch Ness continue and so for the present, and until Bob and his friend Tim Dinsdale get the clear, color, 16mm footage or equally clear 35mm stills that they want of one of the Ness creatures-the energy of the Academy and its members is divided between Loch Ness and the Pacific Northwest.

Both Rines and Dinsdale intend to visit the Pacific Northwest again and, when time from their present activities becomes available, to take part in the continuing and widening search that is centered in The Dalles. Neither Rines nor Dinsdale contribute to the "no place to hide" theory. They have both seen the northwest, have flown over it, and have camped out in its vast forests. They have studied the maps on which we have charted the footprints and sightings. They have examined the evidence and talked with eyewitnesses. They have looked closely at the progress made in our investigations over the last three years and they are satisfied with our work, hoping that a successful conclusion to our slow, methodical research is not too far away.

After those first talks and meetings in Boston in 1971, before Bob and Tim had visited the Pacific Northwest and seen at first hand the magnitude of the vast forests of the coast ranges, their attitude toward the phenomenon of the Bigfoot was one of sensible open-mindedness. Nowadays their points of view have changed considerably. They are still tolerant, but only just so, of people who base their skepticism on the "no place to hide, nothing to eat" theory. And when the non-believer steps forward and drags out the time-worn "well, what about the Loch Ness monsters? No one has ever found one of them," rather than gently chide the unread person for not keeping up with what is going on in the world, they are inclined to smile, turn away, and switch the conversation to other subjects.

One conclusion is that the skeptics are quite wrong. But then, I feel that they deserve to be wrong, for in all cases where their views are adamant-and at times even virulent-they are the views of people who have not and will not take the time to examine the evidence.

Another conclusion is, quite simply, that the evidence that I and my teams have produced does indeed support the existence of the Bigfoot and that there is a small surviving group of them living in the mountains of the Pacific Northwest and British Columbia.

Having come to these deductions-regardless of how unacceptable they may be too many people-there really remains only one more point to discuss. This is the question of what is to be done with creatures and what is to be done about them, now and in the future? The answer to this important question can be divided into four parts.

One, leave them alone. This is undoubtedly the best approach. The record shows, without any doubt, that they are perfectly harmless, that they assiduously avoid man at all times, and that they are totally nonaggressive, shy, gentle creatures that ask only to be left to themselves in the forests. In the history we have studied there is no authentic account of violence toward man and, with the exception of kidnapping-Albert Ostman and Muchalat Harry-there is not even any indication of any interest in man in the northwest the two extraordinary stories of kidnappings-Albert Ostman and Muchalat Harry-there is no evidence of them taking the least interest in man-in us.

Every year hundreds of people go into the coastal ranges and very many families, with children take advantage of the access provided by mountain roads for picnics or overnight camping trips. I think that it is safe to say that people, with their children, have been camping in the northwest forests for probably a hundred years. No child has ever been hurt. No child has ever been kidnapped, or chased, or threatened in any way and, as is obvious to any intelligent person, if the Bigfeet are living in the woods there must have been many times when they have seen children or have been close to them, close enough to capture or even kidnap one had they wished. Perhaps it is their shyness that has prevented them from doing this? Perhaps it is their gentleness? Perhaps disinterest? Whatever it is, the record stands. The Bigfeet are not a threat to man in any way and in the long run, if they could be left alone, if people could just accept their presence and let it go at that, this I feel would be the best policy.

But alas, man in his curiosity has to know more and more about his surroundings and is never satisfied until he feels that he knows all there is to know about it. And with the Bigfoot there is left-until he finds one and can see or touch it for himself-that indefinable aura of mystery that, instead of satisfying man with its intrinsic beauty, seems only to whet his appetite for more knowledge of the subject.

Thus we face the fact that man is not going to leave the Bigfeet alone. He is not going to be content with having them living peacefully in his forests and not bothering anyone. He is going to go after them, seek them out, find out all that he can about them and then, maybe, leave them alone. How is he going to do this? By finding one, by shooting it, by capturing it or by photo-graphing it.

Two, let's shoot one. It seems incredible in this day and age that there are people who would want to shoot something like a Bigfoot. But there are, and I have met them. Not only will they shoot one if they get the chance-heaven hope that they never do-but they even go so far as to propagate this philosophy on the grounds of its justification to science! One individual who actively pursues this murderous approach is basically an example of the other few who follow his footsteps. His proposed *modus operandi,* should the opportunity arise, is to shoot one, cut off its head and then bury the remainder for him to recover later. He would then, he states, show the head as evidence and, having established the actuality of the creature, sell the remainder of the carcass to the highest bidder.

To me, wanting to shoot a Bigfoot is cruel and unjustified thinking. I would never condone it, not even for so-called scientific reasons.

As to wanting to shoot one for monetary gain, this to me is mindless cretinism of the lowest form. The gunmen who follow this path are few and I and my associates keep a careful watch for them in all of our research work. To us they are kin to the people who wound and leave deer in the forest, who shoot coyote and porcupine and badgers for the "fun" of it, who snag salmon and trout at waterfall pools where they lie in wait for their upstream runs, who shoot out of season when young animals and birds need the protection of their parents, who dump old car bodies in the streams and throw beer cans in the rivers and who generally by their intent or their negligence make malignant contribution to the slow but creeping destruction of the wildlife and the environment of one of the most beautiful countries on the earth. Thinking as we do that the creatures are hominid and that they are probably in actual fact a form of sub-man, or undeveloped man, rather than ape, and knowing the feelings of the very many people who now take an interest in the Bigfoot, a protective and almost parental interest, personally I would hate to be the man responsible for the death of one.

Three, capture one. But capture has to be defined. Is it to be permanent capture or temporary capture? Five years of research and many hours of intense discussion on this subject have given me some insights. Permanent capture means just that. It means that the creature or creatures would be incarcerated in a cage and held until it or they died, of disease, of old age, or simply of the psychological effects of captivity. If one is to consider permanent capture one has to consider this side of it. For scientific reasons a permanently held creature would no doubt be of great study value. That its normal life patterns would completely change under activity foreign to it would hardly matter in the light of the amount of material that could be obtained from observation and study, or so it would be believed. The only eventuality that might save it and allow it ever to leave a cage would be if some form of communication were developed between it and its captors that would allow it to promise them that, A, it would behave itself in public, B, make proper use of the toilet, and, C, probably, wear clothes.

Either way, in a cage or in a suite at the Hilton, I feel that the Bigfoot would be as out of place as a wretched, caged gorilla. I would vote against permanent capture and it is not presently, or in the foreseeable future, a part of my planning.

As to temporary capture, it poses many problems. One supposes that one of the creatures would be taken in the forest and then held for a short period during which a team of scientists would be able to examine it, photograph it, et cetera. The problems that would arise here would be the length of time involved in an examination, the means that were used to hold the creature and the temptations that would arise-and that they would there is no doubt-among the examiners not to let it go. For the scientist it would be like allowing a priceless specimen, on which he had worked hard for many years, walk out the door, never to be seen again. And for the laymen present there would be the nagging questions, are we getting all that we can out of this, have we used it for all of its worth? Would it not be worth more to hold it? Or even to kill it and then keep the body? It was George Haas, of Oakland, who summed up what might happen in the case of a temporary capture and who put into words what many of us had thought about over the years when considering this possibility. What would happen, he said, if while you have the creature in hand, either drugged or in a cage, healthy and well, some powerful organization like Disney Studios walks up and offers a check for half a million for the body? Are you then going to let the thing go?

As to actual capture, there are many problems in this area also. In modern terms this means shooting one with a hypodermic containing a serum or drug that will knock it out for time enough to allow a team to get to it, truss it up and get it into a cage. The trouble with this approach is that nothing is known about the creature's blood, its metabolism, its reaction to drugs or anything else. What drug does one use? The common suggestion has been to use one of the serums that have been tried and tested with other primates, and there are two or three of these that are obtainable today with a veterinary prescription. But will they work? Or worse still, will they overwork and kill the target? In the course of our studies we have examined two. One seemed excellent. It was tested on primates and had also been used safely on other animals. It had a fast reaction time and was guaranteed to knock out an elephant, provided that the right dosage was used. To determine the dosage was easy. One simply estimated the weight of the target and regulated the serum accordingly. But how to estimate the weight of something like a

Bigfoot, seldom seen and often varying in size from sighting to sighting? The answer was to use not this drug, but one that had more latitude, a wider safety margin that would allow for error in weight estimate. We found another drug that had these requirements. We could overdose with almost perfect safety. We could dart a 300-pound Bigfoot with a load meant for a 500-pounder and the only difference would be that the creature would stay immobilized longer. But there was a snag. The first drug, the one that required an exact dosage, had a fast reaction time. With large animals it was something close to two minutes and this would be effective as long as the dart was properly placed for an intramuscular injection. This meant that the creature, once hit, would not be able to go very far. The darters would be able to close in at once and get to the creature immediately as it went down. But with the second drug, the reaction time was something close to nine minutes, and this posed very many problems. One was the problem of keeping up with the creature, should it decide to try eluding its pursuers. On top of this was the danger to the creature itself, should it try to scale up or down a cliff, or try to swim a river, or even cross a highway. An agile, fast-moving Bigfoot could go a long way in nine minutes and this possibility presented definite hazards to a potential target.

Leave it alone, shoot one, capture one. Three suggestive answers to the basic problem of assuaging our insatiable past and present curiosity about the creatures. There is a fourth part and that simply consists of photographing one, or more than one, of the creatures and proffering the pictures as evidence. If the pictures were accepted as evidence then little more would be needed as proof of the creature's existence. If they were not then we would be back to square one. If the pictures were not of good quality and if the photographer were not of unquestionably good character, then they would most probably not be believed. For there is no doubt that one can work wonders in a modern studio these days and a clever photographer, with the right props and plenty of time and money, could produce credible looking stills and possibly even movies of half a dozen Big-feet flying over the top of Mt. Rainier, if he wished.

Where pictures are concerned, their acceptance will depend entirely on the credibility of the photographer and on the quality of the pictures. Processing under bond would be essential and the photographer would probably have to swear, before ever even seeing his own pictures, that their subject was a genuine Bigfoot, as he saw it in his lens and as he believed it to be, and that he would stake his own amateur or professional reputation as a photographer and incidentally his own personal reputation, his business reputation, and his social standing and integrity-that the subject was as he stated or at least as he believed it to be. Then, and only then, would he begin to move toward that acceptance that would be essential to using the photographs as positive proof of the Bigfoot reality.

If we include the nearly three years that I spent in full-time investigation of the Bigfoot mystery in the sixties, the total time spent to date is nearly seven years. Moving into the eight year of the long-drawn-out search, I find myself repeatedly faced with the question, is not eight years a rather long time to be spending looking for your hairy friends? Do you not think that you should have been able to find one of them by now? Or that someone else would have photographed one, or shot one, or hit one with a car? Or that some kind of definite proof should have been forthcoming that would have put people's minds at rest about the positive existence of the things?

There are days, truly, when the search seems endless. But when I am looking out the window of my office in The Dalles, watching the evening sun gold the distance hills, or sitting by a campfire deep in the forests of the Cascades with the night birds calling and the stars marching down the sky, I am reminded of men like Dr. Otto Schoetensack, a German paleontologist. In 1887 Dr. Schoetensack became interested in a large sand pit near Mauer, Germany. The pit was close to an area that had yielded much fossil material from the Pleistocene Age and he felt that it would be worth doing some excavating there. He felt this so strongly that he started digging at the pit and did so, continuously, almost every day, for *twenty* years. The result, in 1907, was the famous Heidelberg mandible, one of the great finds of the paleontological fraternity.

There are many other examples such as the decades that it took Louis Leakey of Kenya to find the first real fossil skull, the Zinjanthropus skull, in Olduvai Gorge. And the years that it took to have the gorilla scientifically accepted as a real living creature. First mentioned in European literature in the fifteenth century, the lowland gorilla was not known to science until the late 1800s. The mountain gorilla, a subspecies, was not officially recognized until 1902.

It is obvious to me, as it is to all who have searched for the Bigfeet, that these, the most elusive and man-wary creatures on the face of the earth, are not going to be found within a year or two. This is my opinion and the opinion of the scientists with whom I work, and of my associates at the Bigfoot Information Center in Oregon. But two things could change this rather pessimistic outlook. One would be a positive change in what is presently the biggest need of our search-sufficient funding. For in spite of what some of the glossier magazines have had to report over the last few years, funding of the project has always been extremely limited. (One magazine recently reported that in 1974 the Information Center received a grant of $500,000! We wish that it were true.) Insufficient funding has been the bugbear of The Dalles project, and lack of funds has prevented me and my researchers from putting into action many of the very high potential plans that we have designed over the years, plans that are the end result of thousands of man-hours of study of the phenomenon and of how to get to the bottom of it. Both the International Wildlife Conservation Society and the Academy of Applied Science have been generous in both financial and moral support, as have many private sponsors. But lWCS funds have mostly been applied to conservation projects and, as has already been mentioned, the efforts and funding of the Academy of Applied Science have mainly been used to project the Loch Ness investigations. Full funding of a proper search project has never been available to the present teams and, as a result, the work that has been executed, all of the planning that has taken place, all of the searching that has been done in northern California, Oregon, Washington, and British Columbia, has been restricted and contained to the point where much of it had often to be left unfinished or even, at times of severe financial drought, put on hold.

With enough money to start a full-bodied search and investigation, using the plans that at this time are ready to go into operation, I would predict a definite find, possibly within two years, and certainly within five. The "find" would probably be in the form of first class photographs or 16mm movie film, evidence that in the eyes of some would not be sufficient proof of Bigfoot's existence. But the impeccable credentials of the two institutions now associated with the project would hopefully make it acceptable to science and to the public.

What would the new funding do to the present project and how would it be expanded into a full-scale search? What equipment would be added to the search team's present equipment and what other changes would take place? Briefly, the whole project would be reorganized and expanded. The expansion would include the recruitment of new personnel, among whom would be qualified people in the fields of anthropology, zoology and" biology. They would have to be energetic people, active, dedicated to long hours and hard work and prepared to spend months in the field at a time. Part of their task would be comparative ecological studies in the different areas of evidence, with an eye to determining factors common to each area. These factors could be food, or certain types of tree cover, or even minerals of the type that wild creatures like to eat, such as salt or sulphur veins.

New equipment would include long range two-way radios with a range of several hundred miles and base station in The Dalles at the Information Center. Each vehicle would be equipped with a two-way set for constant communication with base and there would of course be set-calling hours for all personnel. New vehicles would be purchased and this new fleet would be an extension of the present two-vehicle fleet of 197 5 International Scouts, vehicles that have proved their worth in our operations over thousands of miles of very rugged terrain.

Dogs would be used as part of the search team and at this time one of our associates, David Hasinger, of Philadelphia, is exploring this possibility. The plan with dogs would be to train a certain species of dog, possibly a bloodhound or something close to a bloodhound, to follow distinctive scents. Part of the experimentation that is presently being conducted with dogs includes research on scent and as of this year an internationally known laboratory has produced two distinctive scents that contain the glandular secretions of female gorilla and orang, woman, and so forth.

Our newsletter, **THE BIGFOOT NEWS**, would be expanded through advertising and a target of 100,000 monthly issues would be aimed at within two years. Its purpose would be what its main purpose is now, to draw information to the Information Center from the thousands of people who live in the Pacific Northwest, who work in the forests and the mountains or who go into the outback to hike, fish, or camp. Through this huge public relations campaign the eyes and ears of virtually thousands of people would be used by the Center, people who at present are finding footprints or making sightings and are not reporting them to the Center because, A, they do not know that it exists, B, they do not know that their reports will receive serious attention and immediate investigation, or, C, they are not aware of the policy of confidentiality that protects them from the public and from any possible ridicule. The present $1000 reward that is offered for information leading to a find by Center investigators would be increased to, probably, $5000 and through the pages of *The Bigfoot News* this would be made known to the public. With the reward offer would be a description of the policy and aims of the Center towards the Bigfeet, i.e., the total disinterest in a capture and in imprisonment, the care that is constantly taken and that is an integral part of the hunting methods of the center that one of the creatures is not harmed in any way, and the end result of the proof of existence-protective legislation for them both in this country and in Canada.

Many other changes would take place and there would in essence be total overhaul of the present search methods that are so restricted by lack of funds into large-scale expansion and new fields of research and investigation. Lastly, we would begin to move on what we call our geo-time patterns. These GTPs, as they are called, are basically geographic areas where, at certain times, evidence has appeared in the shape of a definite time pattern. For example, Area X. In this area footprints were found in 1953, in 1962, and in 1970. The area also shows four sightings, in 1953, 1967, 1970, 2nd 1971. And the pattern? A study of the calendar shows that all of the footprints were found in the fall months of the year and all of the sightings took place in the same months. This gives us what is called a geo-time pattern for a single area and the result is an area with very high potential for a find. The pattern suggests that in these months one or more of the creatures comes into this particular area. The reasons, as yet, we may not know. Possibly as part of a migration route but more likely to be food. Whatever the reason, the evidence shows some movement and the presence of at least one Bigfoot, for one area, within certain months. This is very valuable information for our searchers, pointing as it does to a single contained area.

But searching one of these areas, or even concentrating searchers in one, can often be a time-consuming and consequently an expensive task. The area may be hundreds of miles from base, and nowadays the expense of reaching into such an area, with men and supplies and all of the logistics involved, can be considerable. To date we have not had the funds necessary to explore thoroughly and take advantage of a single one of the several geo-time patterns that our research has built over the years. When we can, when we start to put into operation the full-scale searching of a fully funded team then I feel that we will at last come to grips with the Bigfeet and with our efforts pierce the shadows of one of the most extraordinary mysteries of our time.

Acknowledgments

(FOR THE 1975 EDITION)

Very many good people have been associated with and have helped the Academy of Applied Sciences-International Wildlife Conservation Society Bigfoot Project over the past five years and have contributed help that ran the gamut from moral support to field assistance to financial sponsorship. Through their interest, advice, and in many cases, actual physical and financial help, the Bigfoot Information Center has been established, the Bigfoot Exhibition has been built and the Bigfoot News, the monthly newsletter of the Center, has been successfully launched. All are to be thanked and I ask to be forgiven if space allows nothing more than a brief mention of their names here.

Among those to whom I am grateful are Anne Adamson, Spike and Red Africa, Len Aiken, Ernie and Dorothy Alameda, Col. Charles Askins, Homer and Phyllis Baker, The Eddie Bauer Company of Seattle for donating some of its excellent camping equipment and cold weather clothing, Al Berry, Herb Beyer, Don Blake, Dr. Geoffrey Bourne, John Bauer of the Oregon Bank, Monte and Sue Bricker, Darrell Buckles, Don and Alta Byington, Bryan D. Byrne, Ken Coon, Jim Craig, Roy Croft of the Skamania County Pioneer, Robert Curley and family, Colin Dangaard of The National Star, Marge Davenport of the Oregon Journal, Norm and Carol Davis, Jim Day, Ian Des Mares, Harvey Dirmuid, Bob Downing of the Akron Beacon Journal, Steve Dunleavy of The National Star, George W. Early, Dr. Cortez Enloe, C. W. (Chuck) Ennis, Dick Enright, Lucius Farish, Phillip Fradkin of The Los Angeles Times, Katherine Freer, John Fuhrmann, Cleve and Manuela Fuller, Dr. and Mrs. Warren Geiger, Bob and Julie Gimlin, Sandy and Linda Golden, Ray and Susie Goolsby, Bill and Beryl Green and Michael Green, Steven Green of the Seattle Post-Intelligencer, John Guernsey of The Portland

Oregonian, Laymond Hardy, Bill and Margo Harper, George Harrison of the National Wildlife Federation, Donna Henderson, Dr. and Mrs. Bernard Heuvelmans, Mr. and Mrs. Robert Hewes, John Hogan of Trail Magazine, the magazine of the International Harvester Company, George and Lo Holton, Jack and Yvonne Hoover, Conrad and Francis Hudnall, Ralph Izzard, Mike Jay, John Jenkins, David Johnson of The National Observer, Mr. and Mrs. George Johnson, David and Carol Jonason, Edward Killam, Vincent and Mary Killeen and family, Dr. Grover Krantz, Mr. and Mrs. Ernie Kuck, Mr. and Mrs. Bernie Kuhn and family, Dick Lacardi of KMED-TV, Medford, Oregon, Peter Light and Linda St. John, Peter Lipsio, Steve and Vera Matthes, Mr. and Mrs. William Marsh, Ed McClarney of the Skamania County Pioneer, Bob and Ardis McClellend, Colleen McKay, Shearn Moody, Steward Mutch, Larry and Ann Nyberg, Mr. and Mrs. Norm Odegaard and family, Mr. and Mrs. William Oden, Ron Olsen, Mr. and Mrs. Dick Park and family, Vern and Barbara Parmentier, Mrs. Pat Patterson, P. J. Reale of the Quincy (Massachusetts) Patriot Ledgers, Dr. Charles Reed, Grant and Loni Robbins, Warren Robinson, of Fort Belvoir, Virginia, Tim and Trish Ryan, Ivan and Sabina Sanderson of S.I.T.U., Dwane Scott, Mr. and Mrs. Ed Sharp, Peter Shepard, Jeffrey Slusher of Fort Belvoir, Virginia, Dick and Edith Sparke, Chan St. Claire, Sheriff of Stevens County, Washington, Mr. and Mrs. Jon Stewart, Nikki Stevens, Roger St. Hilaire, Stanley Stinson, Steve and Shirley Stone of the Captain Whidby Inn, Whidby Island, Washington, John and Delores Suismehil, Al Stump, of the Los Angeles Herald Tribune, Jack and Gail Sullivan. In the city of The Dalles, Oregon, the Chamber of Commerce and its Manager, Bud Hagen; Mayor Donnell and Mrs. Smith and the City Council of The Dalles; Chief of Police Robert Brower and the City Police Department; District Attorney Bernie Smith. Fire Chief Wilson of the City Fire Department, and his staff; Sheriff Ernie Mosier, Sheriff of Wasco County and his department officers, including Deputy Rich Carlson, Deputy Harry Gilpin, and Sgt. Jack Robertson. Father Anthony Terhaar of St. Benedict, Oregon, Bob Walters, Manager of KACI Radio of The Dalles, and his family, Ron and Shirley Ward, Ron and Loren Welch.

A special vote of thanks is owed to certain persons who have generously given more than an ordinary share of help to the Bigfoot project. The time, interest and support that they have provided has enabled us to continue with the project to the present. Among them are: The Directors of the International Wildlife Conservation Society Inc., of Washington, D.C., in particular Leonard A. Fink and Karl Jonas, M.D. Their expressions of confidence in the long-term project are much appreciated. The Explorers Club of New York, for granting recognition to the project in the form of an Explorers Club flag to myself and Russ Kinne, (alias Stuart Mutch) in 1974. Columbia Photo, of Hood River, Oregon and its owners Nicholas and Pam Bielemeier, for much professional photographic advice, work, and equipment and for many overtime hours spent working to assist the project.

For direct support and assistance to the project, the following should know of my appreciation: Frederick Ayer II, Arlie and Polly Bryant, Jack and Ann Bryant, David Chandler, Guy Coheleach, Tim Dinsdale of the Loch Ness Investigation Bureau, John H. Hauberg, Patri Hull, Tom Foley, Bob and Francis Guenette, of Wolper Productions, Los Angeles, California, Nancy Cooke Jackson, Dennis Jenson, Russ and Jane Kinne, Neal and Mary McLanahan, Dr. John Napier, Queen Elizabeth College, University of London, Tom and Mary Page, Ron and Jackie Rosner, Allen Rosse of Washington, D.C. and her son Colin, Marie Roy, Secretary of the Explorers Club, New York, Gerald Russell, Eleanor (Missy) Sabin, Ronald Somers, Lowell Thomas, President of the Explorers Club, Bob and Betty Thomson, of Sausalito and Bronson Trevor of New York.

To the International Harvester Company, makers of the four-wheel-drive Scouts that we have been using through all the years of the Bigfoot project, under the most trying weather conditions of extreme heat and bitter cold, in all kinds of terrain in the rugged coastal mountains of the Pacific Northwest, I can only say this: that their vehicles have never let us down, have given excellent service for thousands of miles, and have inspired our confidence. When a better vehicle appears, we will use it. When it does, it will probably be built by International Harvester. Until then we will continue with the toughest vehicles on the market for the rugged field work in which are involved, International Harvester Scouts.

And lastly, a debt of gratitude is owed to Robert Rines, President of the Academy of Applied Science, and his wife Carol, and to the Directors and Members of the Academy for their continuing support of the project, for most generous assistance with equipment, funding and sponsorship for which I and all of my associates are indeed grateful.

END ... of the 1975 book, *THE SEARCH FOR BIGFOOT*.

Never before published photograph of what is believed to be a genuine Bigfoot footprint, one of several sets of prints found and photographed in September, 1967, by road work crew members, Charlie and Doreen Hooker, in Bluff Creek, not far from the site of the P&G footage but one month before Patterson and Gimlin arrived there. One set indicated a fourteen and a half inch print; the other, above, a fifteen inch print. (The shoe print size is average for a man of medium build, i.e., about 10 inches.) Faint, but just visible, is the strange, (previously referred to by the author) bulge and indentation, behind the big toe. Credit Charlie and Doreen Hooker, 1967.

Rare photograph of ten Bigfoot footprints on a logging road in the Bluff Creek drainage, Six Rivers National Forest, northern California. Photo from author's 60's Bigfoot Research Project; prints found by members of the project team. Records of the size of the prints have been lost, but they were probably the same as what was found in that area several times in the two years and ten months term of the project-fourteen to fifteen inches. Noticeable was the fact, seen many times, of the creatures print's always being on the side of the road, and never in the center. Big cat predators-tiger and leopard-have this same habit, one reason being that it allows them quick access to cover in the event of unwanted discovery by another predator, or by their prey or, in the case of a Bigfoot, one might speculate, by man.

Site of the Bigfoot bed found by the author while in company with the 1960s Bigfoot Project field man Gerry Crew. It was made of moss and was nine feet long, about four feet wide and eight to ten inches thick. The moss had been pulled of nearby trees from a height of twelve feet-higher than the reach of an average man. There no claw or scratch marks on the trees, which there would have been if a bear was involved. The clearing and the nearby area were thoroughly searched for any sign of human occupancy; there was none. No hairs were found in the bed and there was no odor. But in the clearing where the bed was located-seen here-were the skulls and jawbones and rib bones of two deer which had either been killed there or, more likely, killed and carried in there. It was noticeable that the bigger bones were not broken-as they would be if this was a bear kill, to get at the internal marrow-but gnawed by what looked like flat teeth. (See the text for the location of the bed.)

The author's exhibit and information center in The Dalles, OR, in the seventies. Its principal purpose was to have people come in and talk about the phenomenon and, in the process, provide the project with information about sightings and footprint finds. In this respect it was quite successful, part of the reason being that when people who were normally reluctant to talk about an incident for fear of being laughed at (the fear of ridicule) saw that the project was a serious one, they shared and contributed much useful information. In the background is the hill on top of which was built the project's Bigfoot observation post. At first located in The Dalles, the center was later relocated to nearby Hood River. Its creation was financed by the Academy of Applied Science of Boston, MA.

Whistling Ape Ridge, WA, above the Columbia River and opposite the city of Hood River, part of what is believed to have been a one-time north to south Bigfoot route that began at Timothy Lake, south of Mount Hood. Oddly, this name has now been changed to Whistling Ridge. The other landmark in this general area with a Bigfoot connotation, Ape Crossing, on Highway 26, south of Mt. Hood, also –no one seems to know why-seems to have been completely removed from government maps. East of the town of Rhododendron, where Highway 26 crosses Alder Creek, is probably where it used to be located and named.

The author, Bigfoot hunting, at Bluff Creek, Six Rivers National Forest, 1960. He is carrying a 30.06 rifle. In the initial stages of the research, all members of the Bigfoot Project, not knowing what the potential for danger might be with the Sasquatch, carried rifles while in the field; later this practice was abandoned and cameras replaced weapons. Though new to the Pacific North west forests, the author was an experienced big game hunter at this time. However, probably in vanity posing for the photograph, he does not seem have noticed the large crack in the log on which he is standing, indicating that it is going to break and fall into the stream below, quite soon- and possibly he with it.

An unusual photograph of a Bigfoot footprint found by the author in a little pool at the edge of the Bluff Creek stream, 1961. The clear water was about three inches deep; its unruffled, mirror-like surface allowed for this photo. Not having a measuring tape to hand, to gauge the size of the print, the author used a 30.06 cartridge to estimate proportion. The length of the round, according to a sportsman friend of the author, weapons expert Tom (Doc) Donohue, of Pacific City, OR, is 3.340 inches which, by carefully calculated, proportional comparison, made the BF print fourteen inches in length.

This is the precise location of the 1967 Patterson Gimlin footage. The picture was taken by the author in the summer of 1972 and shows the site-now, in 2015, completely obliterated by the weather of 48 years-much as it was the day the footage was acquired. During this visit-one of many to the site-the author used a model, Rick Hodgson, son of a Willow Creek, CA, store keeper, Al Hodgson, in an attempt to obtain an accurate height for the subject of the footage. Young Rick stood 6 feet in height, weighted 156 lbs and had a 32 inch waist. From this, and also using careful measurements of other physical objects seen in the photo-including the tree stumps seen in the background-the author determined that the creature stood 7 feet 3 and a half inches in height and had a 75.36 inch waist. From its physical bulk, as seen in the footage and using his own experience with large Asian wild animals-e.g., buffalo, wild boar, bear and bison-the author estimated the creature's weight as being between 475 and 500 lbs.

This photograph, and on opposite page, are from the Wilson River area, in the coast ranges of Oregon.

Here, on 22 June, 2018, at midday, a most unusual Bigfoot sighting took place. Seven men, all veteran loggers, were working on very noisy, log-loading machinery when they saw a Bigfoot approaching towards them in the area seen here, above the road - where the brush is short - thus allowing them a clear view of the creature. The primate was estimated at about seven feet tall and between 375 and 400 pounds and was dark brown in color. Seeing the men watching it, the primate stopped for a moment and then turned abruptly to its right and walked away and disappeared. The author's interpretation of this very unusual midday sighting is that the creature must have been aroused from diurnal sleep - as well as confused - by the very loud noise of the heavy machinery, which was both powerful and far-reaching.

Close up of sighting area near Wilson River Oregon. Seven Man Sighting area. Note the low growth of the scrub vegetation and the scattered tree stumps, which same allowed for a clear, full-body sighting of the creature by the seven eye-witnesses.

THE HUNT FOR BIGFOOT 2015 EPILOGUE

CHAPTER ONE
Dining at My Table

As I grow older I have become more particular about my choice of companions, especially those with whom I spend my leisure time, whether that be in-house wining and dining, or at a campfire in the woods. The Bigfoot fraternity has among its members not a few witty and intelligent and well-informed people, whose company I have had the pleasure of enjoying across the years. At the same time it also has, like any other unusual phenomenon and as I have already noted, a diverse collection of (to put it kindly) colorful characters. For these people—more often than not individuals whose only subject of dinner table or campfire conversation is themselves and their imagined expertise and (unsupported) findings in the world of Bigfoot—I have no time at all and without hesitation consign them to the category of the uninvited. Among them I list the following.

Anyone who claims multiple sightings and encounters.

Anyone who claims to have seen the BF copulating.

Anyone who claims to have seen them giving birth.

Anyone who claims to have the ability to communicate with them in any way and/or to receive communication in return. (This includes what has come to be known as call blasting; it also incorporates mental telepathy, whistling, singing and the use of musical instruments.)

Anyone who claims to have seen them physically appear and disappear, or grants them the ability to do this.

Anyone who claims that the creatures have extra-sensory capabilities that enable them, among other supposed and extraordinary abilities, to read human minds.

Anyone who claims to be able to find their footprints at any time, at the drop of a hat, so to speak.

Anyone who claims that they steal tools and camping gear at night from campsites. (E.g., one delightful story describes them stealing a power drill and a table saw from the back of a camper's pick-up at night and then coming back the following night to steal the charger.)

Anyone foolish enough to think they can estimate numbers of the creatures based on the present evidence, estimates that range from 100 to the recent statement by a so-called expert (who also claims two encounters) of 100,000 for the whole of the US.

Anyone who claims to have been physically threatened, in any way, by one or more of the creatures, including being charged or stalked by one or more of them.

Anyone who claims to have had one or more living with them as a house guest, albeit in the cellar, or visiting on a regular basis.

Anyone who claims that the creatures can run at high speed, e.g., keeping pace with a vehicle traveling at 30 MPH. (To date, there are no credible accounts of anyone seeing one of the creatures actually running. Perhaps they never run; perhaps, like wild elephants-which are capable of fast walking, but not running-they don't run, simply because they don't have to do so.)

Anyone who claims that the creatures understand and recognize cameras and thus take precautions to avoid same, especially motion sensor cameras.

Anyone who claims that they construct and use shelters and in support of this, produce photographs of brush piles and natural stacks of dead trees.

Anyone who states that they mark their territories by bending small trees over, or snapping them off a few feet above the ground, or by leaving marks and signs on the trunks and bark of trees.

Anyone who believes that their ground sign—i.e., footprints—will stay in place for months on end, so that it can be found again with ease. (A recent release in this regard, for sale by an amateur Bigfoot newsletter scribbler, contained a photograph of what was stated to be two Bigfoot researchers examining footprints of the 1967 footage creature some eight months after they were made—eight months that included a northern California winter, with its rain, flooding, high water in stream beds and river beds, wind, sleet, frost, and snow!)

Anyone who claims to have proof that the creatures make tunnels and use these to travel from place to place, or live underground in caves and / or in the root systems of large trees.

And anyone gullible enough—let's face it, foolish enough—to believe in the absolute and utter gobbledygook of so-called H & I (Habituation and Interaction) sites and their insanely imagined hairy inhabitants.

As I say, I am particular now, in my latter years, about my dinner table and campfire guests, with no tolerance at all for so-called Bigfoot experts, each and every one of whom, it has been my experience, knows close to zero about the phenomenon, their so-called expertise being based on little more than foolish guessing, wild imagination or, in not a few cases, downright fabrication.

I am equally intolerant of the gullible ones, though I have to say I do have a certain sympathy for them, especially those who have been suckered by clever fabrication and fakery into the totally imaginary world of the H & I fantasy, or costly and time consuming investigation into dead areas of zero habitat-a good example of which is northeastern Oregon and southeast Washington-and particularly so when hard cash has changed hands. In regard to H & I research, and the gullibility that supports it, at the time of this writing two of the leading proponents of H & I sites have spent thousands of dollars "developing" them with the aim of making contact with the creatures and, it is supposed, communicating with them. Alas, their huge expenditures—which have, unbelievably, included purchasing properties in areas where the creatures were supposed to be engaged in H & I activities, for use as investigative bases—have produced absolutely nothing, not even a *single* Baby Box Brownie, B&W photograph.

When they are asked why there are no photographs from the H & I sites, the answer from these worthy gentlemen—and also from their perhaps not so worthy salaried H & I site operators—is always the same. The creatures don't like being photographed and make this clear to anyone who goes near the sites. Asked, again, by impudent investigators like the author, how the hairy ones communicate this to ardent photographers—e.g., by word of mouth, or by holding up a sign?—The explanation given is always a little vague. As of now, there are no credible still photographs from the fantasy world of the H & I sites, not one, and the only video footage that has appeared from any of them has been a total fabrication...a character in a fur suit wearing a Star Wars Chewbacca face mask.

CHAPTER TWO
Habitat, Potential Habitat,Empty Zones and Dead Zones.

The question of Bigfoot habitat—where they would live and might be found—is one that is hotly debated within the Bigfoot fraternity. Some of the "experts" in the game state publicly and repeatedly that the creatures are to be found in every state in America, as well as in most of Canada. These claims, because they are supposedly based on responsible research, are actually, unbelievably, accepted by many people. A careful look at them, however, shows they are invariably based on unsupported evidence—vague stories of fleeting figures supposedly seen at night, screams and roars from unknown sources, giant footprints among the daises in Grandma's backyard. In other words, information totally lacking in credibility.

Having invented the descriptive terms, DEAD ZONES and EMPTY ZONES, I should explain exactly what I mean by them. An EMPTY ZONE is one where, for specific reasons, the creatures do not have habitat. These reasons could be the presence of cities, farmland, open and treeless country and areas that have dense human populations. In the case of the Sasquatch, a reasonable generalization would suggest the principal EMPTY ZONE to be anywhere east of the Rockies.

A DEAD ZONE is different. It is an area which, because of its geographical location within the Pacific Northwest and its attractive physical elements—i.e., food, water, cover and space—should have credible evidence of habitat in the form of sightings, footprint finds, recorded early settler history and Native American lore … but does not.

A prime example of this is an area of north eastern Oregon and south eastern Washington-the Wallowa Mountains and the Blue Mountains-which should offer habitat and produce credible evidence of this but, as is discussed below, positively does not do so.

In my own opinion and in that of my research associates, the region that may well be present-day Bigfoot habitat—one that has produced credible encounters and footprint finds, as well having (most important) a solid historical background—is as follows. It is one that has its southern limits in northern California, at a line drawn roughly due east from the coastal city of Eureka. Its northern boundary is the southern border of Alaska. Including the coastal ranges of Oregon and Washington, its western boundary is the Pacific Ocean–i.e., the western slopes of the coast ranges—and its eastern boundary line is, within the US, the eastern slopes of the Cascade Mountains.

In British Columbia, this region commences as a continuation of the Cascade Ranges at the US-Canadian border and then stretches, south to north, for hundreds of miles, all the way to the BC-Alaska border. Its western boundary is the Pacific Ocean and its eastern boundary is a north-to-south line roughly paralleling the coast and approximately fifty miles inland (east) from same.

Within this vast expanse, which of course contains many EMPTY ZONES, Sasquatch habitat seems to be confined to forested regions in mostly mountainous terrain. Much of it has produced believable encounters and credible footprint finds across the years, with only one distinct region being a DEAD ZONE and, as such, failing to produce any credible sign of Sasquatch occupation or movement. This is the above described region of the Blue Mountains and the Wallowa ranges of north eastern Oregon and south eastern Washington.

Bigfoot enthusiasts will be aware that other areas, other states, are claimed as having Bigfoot habitat, among them Arizona, Illinois, Texas, Ohio and Kentucky. However, a careful look at what is claimed by misinformed though often sincere Bigfoot researchers as evidence from these areas, clearly shows that it has no basis in reality and zero credibility, leaving us, with the exception of DEAD ZONES, with what I think is a good, rough rule of thumb for possible habitat…west of the Rockies and nowhere else.

The DEAD ZONE mentioned above, north eastern Oregon and south eastern Washington, is an area that in recent years has produced a number of reports in the form of film footage and supposed footprint finds. However, subsequent, in-depth research by experienced field people proved that *all* of the so-called "findings" were, *without a shadow of a doubt*, 100% fabricated. Further research, some of it by myself, revealed that the area has zero historical background evidence of the creatures, no Native American lore and absolutely no record of them, in writing or otherwise, from its early pioneers, settlers, missionaries, and miners.

FOOTNOTE: This area, a confirmed Dead Zone, is not contained within the Cascade ranges but is included in this discussion of habitat because of the unwarranted and undeserved attention-and subsequent publicity-is has attracted within recent year. The cause of this, as well as its total lack of merit, is discussed hereunder.

CHAPTER THREE

Habituation and Interaction

Having dealt with habitat, I think that it is important that anyone interested in the Sasquatch phenomenon should also take a serious look at something that seems to have emanated in recent years from discussions of habitat and that is now being touted as H & I Bigfoot sites, H & I standing for habituation and interaction.

These are, supposedly, specific, physical sites where, it is claimed, Bigfoot families gather from time to time—hence the word *habituation*—and, while there, allow contact and communication—hence the word *interaction*—with interested members of the human race, namely, us. How the H & I saga got started I do not know but I have talked with a number of people who believe emphatically in its reality or, for ulterior motives, e.g., publicity, or gain, state that they do.

One man–a gentleman of seemingly normal intelligence with whom I talked on the phone—told me that he had purchased a farm in Kentucky in an area where, the farm owners had assured him, there was all kinds of Bigfoot activity, including H & I sites, the farm itself supposedly being one of these. Subsequently, after something like two years of no-show by the local Bigfoots, he sold the farm to a doctor who, with his team, recently (2014) moved in and took over the research.

The first owner told me that he was so impressed with what the locals there told him about Bigfoot presence in and around the area of the farm that he had gone on to support studies of several more—possibly five more—H & I sites in other states and that he was confident that soon—how soon he did not say—they would produce, at the least, footage and still photos of their hairy visitors doing their thing and interacting with the staff he had installed, at least one of whom was a qualified scientist. The only unexpected problem that had arisen, he said, one that seemed common to all of the sites, was that their giant visitors did not like cameras and made it very clear that any attempt to photograph them would be met with instant disappearance and a permanent abandonment of the site.

Asked by me how the hairy ones indicated that they did not like cameras—by word or by deed, maybe by holding up a sign—my informant seemed a little vague. Asked why, after two years, there was still not a single photograph of a single one of them—maybe, for instance, from a camera concealed in an umbrella, or a coonskin cap, or, say, set up in the top of a tree a hundred feet above ground, where the things would not detect it—he did not seem to have an answer. Eventually, as I mentioned, he found a buyer for the principal site, the farm, in this case the doctor who, with a new group of eager beaver BF hunters are now (2015) in place at the site, waiting patiently for the first appearance of its large hairy visitors, to be followed, they earnestly believe, by interaction.

H & I sites. Habituation and Interaction. It would be marvelous if it were true and if a one-hundredth part of it had an iota of reality. But, and I state this emphatically, it is not and I will go on from this statement and yes, at the risk of arousing ire on the part of the H & I believers and supporters, to say that of all of the total and utter balderdash and bunkum and humbug and twaddle that I have heard about the Sasquatch and their alleged activities across the years—which includes their ability to appear and disappear at will, to fly, to tunnel for great distances underground, to walk across the bottoms of lakes— the H & I phenomenon is truly the pinnacle and how anyone can be gullible enough to believe in it, or in the statements of people who claim to have taken part in it—including people who have been employed and paid salaries by sponsors to do this—is completely beyond me.

When I was a small boy, growing up among the delightfully superstitious natives of the Irish countryside, I and many of my youthful companions gave great credence to leprechauns—the little people, we called them—and to fairies, banshees, the tooth fairy, the Easter bunny, Santa Claus, and other delightful fantasies.

But then we grew up.

In summary of the reality—i.e., the authenticity—of H & I sites, hogwash is a good word. Another, from my Irish schoolboy days, is baloney.

CHAPTER FOUR
Food

When I am in conversation with Bigfoot enthusiasts, the subject of food—what the creatures might eat—constantly arises. The more sensible people admit that as of now, we really don't know. We can—as we must do in other areas of our studies of the phenomenon—speculate and when we do this we often come to a mutual agreement, which is that it is reasonable to suppose that the creatures are omnivores and, as such, eat just about anything edible that they can find.

A good animal to compare them with in this respect would be the American Black bear, an omnivore that eats anything that resembles food. All bears—all of the different species—are, to a certain extent, hunters and they will chase and, if they can, kill smaller animals; in the Pacific Northwest this will include all animals, including deer, coyotes, raccoons, porcupines, snakes, ground-living birds and their eggs, rodents, fish, as well as domestic animals, including sheep, goats, cows, ponies, mules, horses, feral dogs, cats, and pigs. They will also— if opportunity arises (I have seen it happen)—kill and eat the young of their own species.

Bears are also carrion eaters and this is an eating habit that enables them to live through lean times such as winter, when deep snow may cover the ground, making hunting and a physical chase difficult. As carrion eaters, they will consume all kinds of food, including dead fish, dead birds, deer and elk carcasses, human remains, and the carcasses of other bears. The amount of decay that has set in, the presence of flies and maggots or the powerful odors emanating from what are often decomposing and putrescent remains, are of no consequence. They will also eat vegetable matter, including grass, shoots, water-plants, berries and fruit, the latter wild or domestic and including apples, often to be found in abandoned orchards. The Sasquatch, as a wild omnivore, would have similar, omnivorous eating habits, ones that would undoubtedly include the consumption of carrion.

Five reliable eyewitnesses with whom I, or friends of mine, have talked stated that they watched—albeit for just a few seconds—Sasquatches in the act of eating. Two of them, professional government surveyors, together said they came across two of the creatures standing waist deep in the water of a small lake, pulling up and eating water lily leaves. A subsequent investigation at the lake in question revealed that its water lilies were indeed edible, bitter if eaten raw—not that a Sasquatch would care about that—and a food reasonably palatable to us if cooked.

The third man, a bear hunter, said that the one he encountered was eating salal leaves—a common bush of the northwest forests—and that it was consuming them by pulling small branches through its mouth, sideways, and stripping them off with its teeth.

The fourth man, a construction engineer, encountered what he described as a young female Bigfoot eating huckleberries, and a fifth man, a Canadian highway maintenance engineer, did not see the one he encountered eating anything; but it was coming out of a canyon where a salmon run was in progress and was carrying a large fish in its hand.

For an omnivore, the Pacific Northwest has ample wild food of all kinds, including many edible plants. It also has large numbers of small animals, including shallow- water fish which a Sasquatch would be able to capture and kill, using its hands. I do not include, within the Bigfoot food chain, the large animals of the northwest forests, such as deer, bear, elk, and mountain lions, because as of now I cannot justify a belief that a Sasquatch might be able to chase and kill something as fast and agile as these animals are. Again, the few stories I have heard of them doing this had little credibility.

At the same time, during my Sasquatch project in the sixties, in northern

California, a bed-like structure that I and a companion—Willow Creek-based Gerry Crew, an employee of my Bigfoot project—found, deep in a remote area, presented an interesting scene and the possibility that I could be quite wrong about the ability of the creatures to chase and kill large annals. The bed—if this is what it was—measured nine feet long, four feet wide, and about ten inches thick. It was made of moss, which appeared to have been hand-stripped off nearby trees up to a height of twelve feet. The complete absence of evidence of human occupation of any kind—paper, plastic, shell casings, fishing line, toilet paper, cigarette butts—plus the blurred remains of a single, large, five-toed footprint in nearby wet sand—convinced us that what we had found was not man made, but very possibly the work of a Sasquatch.

Of even greater interest, however, was what we found lying in the clearing around the bed, which was the remains of two deer, bones and skulls, which find leaves us with a question mark concerning the ability of the creatures to chase and kill large animals and whether these are part of their food chain.

CHAPTER FIVE
Goo

As many of my readers will know, I have spent quite a few years living in Asia and as a result have become proficient in some of the languages, particularly those of India and Nepal. From both of these countries, across the course of time, many native words have been carried back to the West, in the case of India mainly by soldiers of the British army. Some, unnoticed, are actually in use today in the English language, such as the Hindustani words for a ball-*gouli*-and for a clothes closet-*almirah*-and for curry-*masala*. One that is not so well know-and not quite in use in the West- is goo. It is a Nepalese word and it stands for wildlife fecal matter, droppings, scat, etc. It has a certain ring to it that I like and so I am using it in title to little chapter, which is a brief discussion of Bigfoot feces and where we stand with its lack as well as—via a story told to me by a very reliable person-quite possibly why none has been found to date.

Several of the books of recent years written about the Bigfoot phenomenon talk about the finding of Bigfoot goo. There have also been many papers and many claims of its being found and being examined, with in-depth details of its contents. The fact is, no one has ever found any fecal matter, of any kind, that has been positively certified as having come from a Bigfoot.

My guess is that in physical form, Sasquatch droppings would be similar to ours. In other words, cylindrical in shape and sausage-like in general structure. One Pacific Northwest animal that produces goo in this shape is our Black bear and I cannot count the number of plies of same that I have seen in the north west forests since I first came here. I have examined some of it from time to time and its contents has always clearly indicated its owner to have been what we know him to be to-an omnivore with a diet that is all inclusive and *goo* that contains everything from berries to grass to bird feathers and the bones of fish and many smaller animals. The question in my mind of course, with the large numbers of piles that I have seen and examined-if only cursorily-is, could one or more of the piles that I saw have been from a Bigfoot? I don't think so, in that, unless it was a young Bigfoot, I would guess that the big primate's droppings would be much larger than those of the average Black bear. The contents of course would be similar, if the Sasquatch are omnivores, which I personally believe they are.

So why have no Bigfoot droppings ever being found? There are two answers and the first is that they have been found and possibly even collected without the collectors knowing what they had and with an aftermath that did not include scientific examination.

Scientific examination, if applied, would be twofold. One, an examination of the fecal contents to determine if there were parasites that were definitely not found in bears or other northwest animals. Two, and more important, an examination of the outside surfaces of the droppings for natural, anal lubricating oils that could contain DNA.

Some years ago I guided a British team studying wild elephants in the south west Nepal Terai forests. I was worried, before they arrived, about the darting of the big pachyderms that would be involved because no matter how carefully this is carried out, there is always the danger of casualties, either from overdose or from physical injury when the big animals succumb to the drugs and go down. So when the team arrived I was delighted to find that elephant darting-for the purpose of DNA collection-had become thing of the past and that the new method was simply to follow the animals, get close and collect fresh goo samples, the outside surfaces of which would carry anal canal, lubricating substances that contained DNA. For a Bigfoot, if and when we find their goo, I believe the same procedure would apply.

The second answer to the lack of Sasquatch goo is that the clever fellows hide it. Possible? Yes, definitely possible. Probable? Maybe a little farfetched but yes, probable, just as is the theory that they hide their footprints.

What comes to mind immediately in the "hide the goo" hypothesis is, are there are other animals that do this? The answer is yes, even if it is not thoroughly done much of the time. Both tigers and leopards make an effort to hide their droppings before and after fecal evacuation, scratching the ground vigorously to start-to make a shallow hollow for the goo-and then doing the same afterwards, to at least partially conceal it. Our domestic dog does the same thing, even if in a feeble way, with usually nothing more than a couple of quick paw thrusts on departure. This is a very ancient habit with *canis domesticus*, but it is still instinctually there.

But a Sasquatch? The answer is that I don't know. Nor does anyone else and with one exception-a short story that bears telling here-there is no record anywhere, ever, of anyone actually seeing a Sasquatch practicing physical evacuation. The story? It's an interesting one and I leave its interpretation and credibility rating entirely to my readers.

Among the interesting people whom I met when I first came on the Bigfoot scene in northern California, in 1960, were Ernie and Dorothy Alameda, of Hoopa, California. Both in their early fifties at that time, they owned the Oaks Café in the Hoopa Native American Reservation, about halfway between Willow Creek and Weitchpec, and soon after I arrived in the area I made their acquaintance over an early morning breakfast of eggs and hotcakes, huge hotcakes, designed, as Ernie put it, for logger's appetites. After a number of visits we became friends and began dining together regularly at their home-a very pleasantly appointed apartment above the café-during which I soon discovered that they both had a serious and intelligent interest in the Bigfoot phenomenon.

Ernie Alameda was a genial but tough, no nonsense Portagee American-as he called himself-from Oakland, with a colorful background that included everything from being a ride-the-rails hobo, to a street fighter, to a firewood seller-driftwood hooked out of the passing current of Oakland Bay with a wire hook and carried on his back and sold as kindling in the streets of San Francisco-to a bar bouncer and short order cook, to what he had become at last, a much-liked and well respected café owner on the Hoopa Indian reservation. Dorothy was a Yurok Native American and was one of the most gracious and charming and delightful woman I have ever had the privilege of knowing. She was, with her quiet voice and gentle manners, in many ways the exact opposite of Ernie. But somehow or other this seemed to help shape what was obviously a delightful, loving and enjoyable relationship.

Dorothy told me several stories about the Sasquatch-or the *omah*-which is what she called the creatures in her own language. She told me that the Hoopa Indians believed a group of them lived in a deep, densely forested, nameless canyon west of the reservation, one through which a stream called Mawah Creek flowed south west, to empty into the Klamath River, south of Weitchpec. She said that none of the tribe ever went in there, even to hunt with rifles. She also told me about an encounter her father, now deceased, had with an *omah*, many years earlier.

He had gone to collect crawdads-fresh water crayfish-in a canyon called Dark Canyon-a shallow ravine that dropped into the Trinity River south of Willow Creek-and on this occasion had ridden there on horseback. Tethering his horse at the mouth of the canyon he set off upstream, on foot, with his fishing basket, to where he knew his the little mudbugs-as they are also called-were plentiful and, arriving there, walked out into the middle of the shallow water, squatted down and began carefully overturning the bottom rocks under which the little crustaceans made their habitat.

The streambed, Dorothy's father told her, was composed of a long, descending tumble of rocks of all sizes, some of them quite large, with fast-running, shallow water cascading down over them and her father had been there just a short time, finding his little aquatic specimens and dropping them into his fishing basket, when his eye caught a movement about fifty feet away, upstream from where he was squatting. The movement was an *omah*, coming out of the forest at the edge of the stream.

Seeing the huge primate, her father, already squatting, lowered his profile another two or three inches and kept perfectly still. The *omah*, he told her, took one step out of the wall of brush that edged the water, looked up and down the stream and, not noticing the crouching man, walked out to the middle of the streambed. There it climbed on to a big rock, squatted down, turned its body around so that its buttocks protruded over the edge of the rock and defecated into the fast running water. This done, it stood up and for a moment stood and stared upstream, away from the crouching man. Then it slowed turned its body right around and looked directly at him and their eyes locked together.

For a brief few seconds they stared at each other. Then the huge creature swung its body around, stepped off the rock into the running water and with a couple of long strides was into the streamside brush and gone.

Although it has no real bearing on the "hide the *goo*" hypothesis, it is interesting to note what Dorothy's father had to say about the creature turning to look at him like this. He said that he had not moved a muscle from the minute the *omah* came out of the brush, nor had he made any sound -apart from which any small sounds he might have made would have been covered by the noise of the fast-running water. And so it seemed to him as though the creature had actually sensed his being there.

If this story is true-and I have no reason to believe otherwise-and if the Sasquatch in question was doing something that it did every day, i.e., dropping its feces into moving water, then what we have is two conclusions. One, that it was doing this to deliberately conceal its droppings and, two, that it was doing something that many if not all of its giant companions naturally did, all of their lives.

One day someone will see a Sasquatch actually, physically, depositing its goo where it can be found and collected. Until that time, in spite of all of the wild claims that are being made of findings-which same will continue ad infinitum-we are and will remain sans *goo*, like it or not.

FOOTNOTE. The story that Dorothy Alameda's father told her about his encounter with an omah, in Dark Canyon, appears in the author's 1975 book, THE SEARCH FOR BIGFOOT, on Page 138. There it is anecdotal and simply included as part of the author's research of that time. Here, in THE HUNT FOR BIGFOOT, it is included as an integral part of the chapter on Bigfoot goo and the question of why it is never found or, at least, why it has not been found until now.

CHAPTER SIX
Sounds

Among Bigfooters, as we call ourselves, endless discussion revolves around what kind of sounds a Sasquatch might make. People claim they have heard—out there in the boonies, as the saying goes—roars, screams, howls, yells, whistling, screeches, squeals, moans, groans, grunts and growls and, recently (2015), from a visiting Bigfooter, didgeridoo-like whirring noises, all of which they attribute to the Sasquatch.

My personal opinion, the same supported by not a few of my associates, is that the creatures are totally silent—something that in common sense one would expect from a shy and elusive primate that obviously spends much of its time avoiding man; that is to say, that they make no sounds at all and that the noises that people are hearing—and quite honestly though mistakenly attributing to the creatures—all have natural sources in the wildlife of the northwest forests. I have heard many of these sounds myself and have always been able to identify the sources, among them coyotes, foxes, ravens, crows, owls, deer, elk, wild turkey, feral peafowl, mountain lion, and bear.

One sound that bothered me and for a little while, when I first heard it, was the above-mentioned didgeridoo-like whirring, a sound normally associated with a musical instrument used by Australian aborigines when dancing, made by a length of stick shaped like a blade, with a rope handle, an instrument that is whirled over the head at speed. I was hiking alone, on one of the ridges coming off Mount Hood, in northern Oregon. It was evening and the sun had just set and the sound—a deep whirring drone—started up in the forest just ahead of me. I froze for about thirty seconds and then, to identify its source, decided to stalk it.

I did this, using the skills I have developed across the years in stalking tiger and other large wild animals, by moving very slowly from tree to tree, endeavoring, all the time, for the sake of concealment, to keep one tree trunk between me and the source of the sound. When I reached what appeared to be the last tree, with the sound coming from the other side of it, I put part of my face and just one eye around it very slowly, to see the source of the sounds… a large owl, sitting on a stump, staring back at me with big brown eyes.

Loud whistling has been reported by hikers in the northwest forests, but always without the actual source of it being identified. Of interest in these claims, however, is a report by a missionary, the Reverend Elkanah Walker, writing about the Colville Indians in 1840, including in his writing a mention of what he called "a race of giants" and stating that the Indians said they always knew when "they" were around because of the whistling sounds they made. But, as I say, we have yet to have a credible report of someone actually seeing and at the same time hearing one whistling.

Again, if you spend enough time in the northwest forests you may eventually get to hear the powerful, incredibly loud roars that people hear from time to time and are attributed to the Sasquatch. I have heard them, as have others, some of whom have been frightened by them.

My experience with the roars in question took place on a summer's night in the Hood River Valley, many years ago. I was sleeping on the second floor of my house and, as it was a warm night, I had my bedroom windows open. The house was built on a hill and my land sloped down, east, to the East Fork of the Hood River. From there, where it became part of the Mount Hood National Forest, it rose steeply through thick forest all the way—about a mile—to a jagged rock escarpment. The sound, which emanated from the forest below the base of the escarpment, consisted of two sets of powerful roars, each lasting about five seconds; because of their power and intensity in the clear night air, they sounded quite close.

The first set jerked me right out of my sleep and the second had me out of bed and leaning out the window. After the second set there was silence.

I have talked to three people who have heard these incredible roaring sounds. One was Bill Munroe, an experienced outdoors and nature writer for a Portland-based Oregonian newspaper. The sounds, coming out of a thicket about a hundred yards in front of him, made him nervous to the point where he returned to his pickup and got inside and closed the doors. He did not see what made the sounds and was skeptical that they might have been made by a Bigfoot. But he did contact me to ask me what I thought about them and we had a pleasant lunch together, courtesy of *The Oregonian*.

The second man was raccoon hunting—illegally—at night with a small pack of dogs and when something began roaring in the woods nearby, his dogs, he told me, turned and fled. He was quite convinced the sounds were made by a Bigfoot. They were very scary, he said, persuading him to abandon the hunt and follow his dogs and leave the area, at speed.

During my nineties Sasquatch project, BIGFOOT RESEARCH PROJECT 111, I met a fourth man who said he had heard the same roars and, as I had heard them myself, I knew that, whatever made them, they were real, and not imagined. My curiosity was aroused to the point where I delegated one of assistants to investigate them. I told him to start by collecting recordings of the sounds made by all of the large animals of the Pacific Northwest, to wit, mountain lion, bear, deer, elk, and feral wild boars. Within a week, we had what we wanted, a clear, professionally made recording of the roars that I had heard and that several others, when I played it for them, positively identified as what they also heard. The source...a hungry Black bear in the San Diego Zoo.

I do not in any way belittle people who tell me about hearing what they think were Bigfoot sounds, roars, screams, howls, etc. With some exceptions they were not being untruthful and I believe that they did hear the sounds that they described. Their mistake was in not being able to identify them and therefore attributing them to a Sasquatch.

CHAPTER SEVEN
Intelligence

As with numbers, habitat, etc., the most absurd theories are put forward by amateurs in regard to the level of intelligence of the creatures. One of these is that they obviously have a high level of intelligence, clearly indicated by their ability to avoid man with the success that they do. However, it is my opinion that, with the rare exception of skilled woodmen and experienced hunters, the average person is as easy to avoid in the woods as a logging truck grinding up a steep hill in first gear.

A good example of what I am discussing here would be one man with whom I went out, some years back, an older gentleman who was a retired US Fish and Wildlife employee. Because of his background—which included many years of Pacific Northwest fieldwork—I had looked forward to spending a few hours in the woods with him and hopefully learning something new from his expertise. Alas for my hopes, for on arrival on our chosen site he got out of my car, slammed the door with a bang—the sound of which, I would guess, went half a mile—lit up a cigar, turned to me and shouted, through pungent clouds of cigar smoke and with a voice that could easily have carried four hundred yards, "NICE DAY, WHAT!"

Another trio of so-called woodsmen with whom I had the misfortune to be in company for an afternoon, spent all the time shouting to each other in loud voices. When I asked why they did this they looked at me as though I was a little slow. "To keep in touch" was their answer; too easy to get separated out here, you know!

I have spent many years studying and writing about animals (six books in print), mostly the wildlife of Southeast Asia, with the emphasis on large animals such as tiger, elephant, rhino, buffalo, and bison. In my early years and when I first became aware of their ability to avoid me, seemingly to sense my presence, I put this down to a high degree of intelligence. Then, as time and fieldwork taught me, I saw that what I was encountering was not animal brainpower, but, simply, animal instinct, with alertness and awareness based on this.

A tiger, for instance, coming in to its kill to eat and being shot at by a hunter from a tree platform will—wounded or not—never again allow this to happen and after one experience will approach the area with caution that will include intense visual inspection of all the nearby trees, a total change of behavior from its normal one of never looking above the level of its eyes. Some people might regard this as intelligence; I would classify it as a lesson learned and instinctively ingrained.

If the creatures of our interest have an enemy, it is almost certainly man. The big wild creatures of our northwest forests—bear, deer, elk, wild boar, and mountain lion—are of no danger to them and, as such, of no consequence other than commonsense avoidance. We don't know—there are no records—what lies in the brain of a Sasquatch about what kind of danger we represent to them. But there undoubtedly is something there and my guess is that at the merest suggestion of the presence of a man—banging car doors, cigar and cigarette smoke, loud or even distant voices—their instant reaction is flight and concealment, something that does not require any more intelligence than that of a deer or a bear.

The only thing that I have encountered that suggests intelligence a little superior to that of any of our common Pacific Northwest animals is a set of finds I made during my nineties Bigfoot research project. These are listed in my journals and records of that time as Rear-View Mirror Sightings.

If you live in the Pacific Northwest and drive a car, you will, across the course of the years, have seen many dead animals on the roads. Possibly bear and deer. Maybe coyote and birds. But, more often, raccoon and possum. I have not personally kept count, but I have probably seen thirty or forty of these little animals—road kills they are called—in the years I have lived here.

A Sasquatch, living in the Northwest and traveling from place to place, north to south or east to west, has got to cross public roads from time to time, so that one of my thoughts concerning an eventual finding, for many years, was that eventually one would get hit by a car or truck. That this has not happened to date may indicate a degree of intelligence advanced enough to make them aware of the dangers of fast-moving traffic, instilling in them a degree of caution which none of our other forest-dwelling creatures seems to possess.

Three high-credibility eyewitness reports, unearthed during my last project, were of the creatures being seen by drivers in their rear-view mirrors, crossing the road behind them.

One man was a school teacher returning from work to his home in southern Washington and I found his encounter unusually interesting. He had driven the road for many ears and knew it well and on this particular day, on a straight stretch with light bush on either side, something caught his eye that was unusual, something that he did not recall having seen there previously. It looked just like a big stump and the reason it attracted his attention was his familiarity with the particular stretch, and his awareness that in the many, many times he had driven it, to and from his work place, he had not noticed it there. This aroused his curiosity so that he took his eyes off the highway for a moment to look into the rear-view mirror, if only to make sure that he had actually seen it. And what he saw was the "stump" rise, stand up, and in three quick strides cross the road behind him.

If Sasquatches have this cerebral ability, one that places them above the animals that are perennially hit and killed by cars and trucks, then I will allow that they do show a degree of intelligence somewhat higher than that of their woodland companions. Beyond that? Beyond that I do not know, never having seen or heard anything credible in connection with the creatures to indicate anything approaching superior intelligence, or human intelligence. Nature endows all of us—man and animal—with everything we need to survive, including thought not limited to camouflage, brains, instinct, physical ability. What it does not give us in this respect is what we do not need and in the case of the Sasquatch I personally think that it has all that it needs in the limited intelligence that it seems to possess. Certainly enough to be able to survive very well for a very long time, not the least of this ability being clearly indicated by its continued success in avoiding man.

CHAPTER EIGHT

Ground Sign, Footprints, Structures, Tools.

Skeptics of the phenomenon constantly seize upon the lack of physical sign left behind by the creatures as an indication of its unreality and it is true that very little physical sign is ever found. One pleasant little equation they smilingly offer is what ten Sasquatches should leave behind, in the form of sign, on the ground, if they are walking and leaving footprints and, with an average stride of 40 inches for each individual, together walk ten miles. The answer is 15, 840 footprints. The equation sounds unreasonable; but there may be an explanation for it.

Firstly, of course, all indications are that the creatures do not live in groups, as a result of which there is very little possibility of our encountering a group of ten or more. Much more likely, if we find sign, it will be of one, or at the most two. If it is one, then the creature, traveling a mile and using the 40-inch stride figure, will place its feet on the ground 1584 times. If there are two that figure will double, to 3168. So what is the answer when even a traveling twosome should leave as many footprints as this, and we are not finding them?

For skeptics, the answer is, simply, that they do not exist. But for those of us who think they do, the explanation may well lie in the composition of the surface of the ground where they live, which, more often than not, is too inflexible to register footprints, an example of this being the hard surfaces of the thousands of miles of gravel surfaced logging roads, which they may well use—nocturnally at least—for travel. Another example is the vast, wild areas between these roads, where surfaces are covered with leaves, twigs, thick grass—both fresh and dried—broken sticks and very often thick carpets of pine needles, which actually spring back to reform their surfaces within minutes of being compressed. Soft dirt, mud, shingle, sand and snow will all take and hold physical imprints for a short time. But then, bowing to the vagaries of Pacific Northwest weather—rain, wind, snow, sleet, frost and defrost—they soon regain their original surface compositions and whatever was briefly there is permanently gone.

There is another answer to the lack of footprint finds and I have to admit that it is a little far-fetched. This is that the creatures know that their big feet leave sign on the ground—sign which a man might follow—and, whenever they can, choose ground that will register as little sign of their passage as possible. This, going back to the unknown area of the creature's intellectual capability, indicates a level of intelligence not found in wild animals. But it is possible and if the practice is followed by them, then it can be successfully performed for miles. As an experiment, I have done it myself and some years back, together with my friend and fellow Bigfoot enthusiast Dennis Jensen, I walked through a mile of high mountain forest without leaving a single print on the ground. For our little experiment we used patches of logging road gravel, ground thick with stick debris, rocks, stones, moss, logs and carpets of pine needles leaving behind, we hoped, little more than "a grain of sand upon a leaf" making it extremely difficult, if not impossible, for anyone to follow us.

As to tools, and the possibility of the creatures having them and using them, again we have no credible evidence. I do like the fairy tale stories of them stealing tools from campsites at night, especially the one where a power drill was taken one night to be followed, the second night, by the creatures coming back to steal the charger. But, seriously, there is no credible evidence in my research, and that of my associates, of the creatures using tools, or even being seen with any kind of tool—not even a stick. In fact, the only credible report that I have in my records of one being seen handling and carrying anything comes from BC, many years ago, when one was encountered by a Canadian engineer from Pemberton, walking down the side of the road carrying a fish.

As to structures, just as stone walls do not a prison make, nor iron bars a cage, so piles of brush, heaps of fallen logs, bundles of sticks and grass, do not, as is often claimed by amateur naturalists, a Sasquatch dwelling make.

As to their deliberately leaving sign—either of their passage or maybe as a form of communication with others of their group—again there is no evidence. One natural phenomenon constantly pointed out to me by newcomers to the scene as being Bigfoot sign is bent trees, i.e., small trees bent and twisted out of their natural shape. "Ah," they say, "Bigfoot was here; that's his sign." But the answer to this is very simple; the bending is often caused by another tree falling on the one in question, bending it down and then slowly slipping away to leave the bent one to recover on its own. Winter storm damage is part of the answer to this little puzzle.

More often than this, however, the reason is winter icing, where thick ice will take the sturdiest of young trees and, with its sheer weight, bend it double. It used to happen to my blueberry bushes when I owned and ran an organic, one thousand-bush blueberry orchard in the upper Hood River Valley, in northern Oregon. At the end of winter, getting the bent bushes—and other small trees on my property that had suffered a similar fate—back to their normal and upright shape required the generous use of vertical poles and bungee cords.

CHAPTER NINE
Personal Experience and Contact

People who claim sightings of the creatures (real or imagined) express amazement when, after they ask me about my own experiences across my five and a half decades of research, I tell them that apart from footprint finds, and what was very possibly a Bigfoot bed, I have made no significant discoveries, have never seen one, never heard one, never verified any scat, never even found as much as a single hair. Two "experts" with whom I have met recently promised me that if I came out in the woods with them they would quickly show me footprints and if we went at night we could listen to the creatures chattering back and forth, whistling, laughing, etc. Maybe I have reached a point in my life where I am just too cynical to accept what, to me, are quite ridiculous proposals? Maybe I am missing something by not going out with them? Or maybe my own years in the field simply tell me that their promises are as empty as your standby, emergency, empty, five-gallon gas can when you are out of gas and stranded … fifteen miles back in the boonies.

So then, what do I have to offer, after fifty-five years of chasing Mr. B? Very simply, not a great deal. Firstly of course are the footprint findings that we found in northern California during the sixties project and also the phenomenon's extensive history and background, that our dedicated researchers unearthed. Then the above-mentioned finding of what we believe was a Bigfoot bed. And lastly, an experience with sounds that I am unable to explain. I will try to tell both stories of the bed, and the sounds, here, now, starting with the story of finding the bed.

The place where Gerry Crew and I found the bed was in a remote area, the location of which, when we set out to search it for footprints or other sign, only my staff knew. (Keeping details of our search locations confidential was important; it helped to preclude the possibility of hoaxers getting in ahead of us and fabricating sign, on which a lot of time could be wasted in investigation and follow-up.)

The location of the find was halfway down a creek about twenty-five miles in length and access to it was by wading the creek for many miles, knee-deep in ice-cold water and through mountains of wet and slippery debris. The creek had a road paralleling it, high up on its south side, about 1000 feet above it, one that ran all the way to its source in a watershed; Crew and I used this to get to its source, being dropped off there by Steve Matthes, a member of my research team.

We planned to do the hike down the creek in two days, covering about twelve and a half miles a day and by late afternoon of the first day were close to halfway down. I was leading and wading in the middle of the creek when I noticed something yellow, up on the creek's ten-foot-high, left-hand bank.

FOOTNOTE: An important part of our fieldwork in those days was searching for footprints or other sign and we did this in areas the ground surfaces of which gave us the best opportunities. This often meant streams beds, where there were soft surfaces, such as sand and mud. We also used the open courses of stream beds to hike in and out of areas of dense forest, especially if there was near-impenetrable deciduous growth. The going was often cold and wet and invariably included clambering through or over huge piles of debris and fallen trees. The alternative, logging roads, made for faster travel, by car or on foot. But their surfaces were invariably hard, of crushed stone, or gravel, on which most animals left hardly any sign.

I asked Gerry to wait a moment and climbed up the bank to have a look, to find the aforementioned bed set in the middle of a small clearing. The bright yellow, which had caught my eye, was the color of the decaying moss of which the bed was made and which, as I have stated already, we found had been pulled off nearby trees to a height of about twelve feet, a level higher than either of us could reach without a ladder.

I walked back to the stream to get my companion and that was when we both noticed a single, blurred, five-toed footprint in the wet sand at the water's edge. It was about fourteen inches in length and appeared to be about a day old.

By now, late afternoon, in the deep gorge of the stream the light was fading and as I wanted to carefully examine the bed in good light, we decided to camp for the night and do this in better light, in the morning. We found a flat, clean area at the edge of the clearing and got a fire going and heated up a couple of packets of dried beef stew. We then gathered some leaves and made ourselves a couple of rough beds and spread our sleeping bags on them and settled down for the night. At least I did. Crew was distinctly nervous about the possibly of the bed owner being in the vicinity and resenting our presence, so during the night got up half a dozen times to keep the fire going. In the morning my sleeping bag was covered with a thin layer of white wood ash.

We were up with the dawn and got some coffee brewing and then set about examining the bed. We searched carefully for hairs but found none. We also sniffed the bed for odors and found it to be clean and without any scent. We then examined the bones that lay all around the clearing. There were two skulls and many pieces of rib and leg bones. One thing noticeable was that the bones had been stripped of meat and tissue without any grooving—in other words, by what must have been flat-edged teeth. Another thing we noticed, keeping in mind the possibility of bear, was that many of the bones were still intact with their marrow contents untouched inside; bears, with their powerful jaws, will always chew up bones like these to get at the marrow.

Lastly, again keeping bears in mind, we carefully examined the surfaces of the trees from which the moss had been stripped, looking for bear sign. The Black bear is an agile climber and will clamber straight up the smooth, limbless surface of a tree, using its claws for traction. But when it does, it will leave distinctive scratch marks on the tree's surface, short ones when going up and longer ones when coming down. But there was no sign of grooving or scratch marks of any kind on the surfaces of the trunks, thus, we felt, precluding the possibility of bear and leaving us with the conclusion that whatever had stripped off the moss had done so manually.

We photographed the bed and the bones and the clearing, using Kodachrome color film, and then set out, on down the stream, another twelve or so miles, to its confluence with the Klamath River, which we reached in the late afternoon and from where we were picked up by one of my research volunteers and driven back to our base at Salyer.

The photographs that Gerry and I took were all subsequently lost. However, on a second trip into the site, with Tom Slick and his brother-in-law, Lou Moorman, and accompanied by Jim Crew, Gerry's nephew—also working for me at that time—and Steve Matthes, we took more photographs and two of these—in which the deer bones are visible—appear in this book.

For this second trip we accessed the area by Jeep on the same road that Gerry and I had previously driven to get to the head of the creek, one that paralleled the course of the stream and ran on a ridge high above the canyon's south side. This time we stopped about twelve and a half miles in and got to the site by scrambling straight down the very steep, brush-covered, thousand-foot wall of the canyon.

This second trip was made within a few days of the find by Crew and myself. We found the site undisturbed. The footprint in the wet sand at the stream edge was gone, washed away, and we found no others.

For the interest of readers, the name of the creek where Gerry Crew and I made our find was Red Cap Creek and it is located in the Six Rivers National Forest, in northern California. Its confluence with the Klamath River is a few miles south of the little settlement of Orleans, on Highway 96, and if memory serves me well enough after the more than five decades that have passed since our find, the road that we used to access it, running north east off Highway 96, was called the Shelton Butte Rd.

As to my experience with sounds, it is one, I will admit, that is subject to ongoing interpretation and analysis, which I will allow my readers in that, to this day, I am not quite sure of the source of the sounds I heard. Nevertheless, as it is one of the only two times in the years of my Bigfoot research that I may have come close to one, even close to seeing one, I believe that is worth including here. The story is as follows.

In *The Search for Bigfoot*, my 1975 book which is contained within this new edition, the reader will find a detailed account of a dual sighting involving three men. The men's names were J.C. Rourke, Firman Osbourne, and Jack Cochran; all three lived and worked in the Hood River Valley, on the Columbia River, in northern Oregon, on the edge of which the sighting took place. The location was the top of a big hill called Fir Mountain, a small, approximately 2000-foot-high prominence that is part of the great ridge that encloses the Hood River Valley on its east side and the sightings took place here in June, 1974. A detailed account is to be found in the chapter entitled A LOOK AT THE PRESENT.

Not included in the account is an important part of my subsequent investigation. It should have been contained in the original account, or at least attached to it, but I sent it to the '75 book's publisher and he failed to include it, stating that the book was getting too big and that it did not need any more data than what it already had.

The three men involved in the incident had their separate sightings spread across two days, and, as I mentioned in the '75 account of the incident, I was lucky enough to pick up the story and meet with them—and actually be taken to the site by them—within just a few days of the dual incidents, while there was still fresh sign on the ground and, I thought, while there was still a possibility that the creature they saw might be in the area.

Wondering what I could do as a follow-up, I studied the notes that I had taken when talking with the men and from them tried to formulate a plan. One thought I had was of setting up an observation post—a site from which the location of the sightings was visible—on a hill near the location of the sightings; I could keep watch, using high-powered binoculars, through the daylight hours, of the whole area. Another was to drive the back roads of the ridge, day and night, in the hope of a sighting, or of finding more footprints. Both plans were feasible and seemed sensible and I was tossing them back and forth when a question came to mind: What was the creature actually doing when the men encountered it? Walking through the forest? Sleeping at the edge of the clear-cut? Maybe digging for roots?

FOOTNOTE. *Getting information on a sighting this quickly was and still is quite rare. More often a report is months or even years old. The dual sightings in this case—according to my journal of that time— took place on Thursday, June 20th, 1974–Cochran—and Friday, June 21th—Osbourne and Rourke—and I received and acted on the information on Saturday, June 24th.*

The answer, of course, was, none of the above. The answer was that it was watching them, watching them working in a clear-cut, i.e., driving a small tractor, stacking slash (brush, logging debris) preparatory to burning it later. And if it was watching them, what did this suggest? It suggested, to me, the possibility of a sense of curiosity. If this was true, then a plan for further investigation—an ongoing working plan—seemed to lie in creating something that might again arouse its curiosity and bring it back again to the clear-cut, within visible range, within camera range. This would require something better than an observation post or twelve hours of road running. An "attractant" of some kind, as I called it in my journal.

With this plan in mind, and after meeting with the three eyewitnesses and visiting the site with them, on the 24th, I drove back up to the top of the mountain the next day and set up a little camp, right in the middle of the clear-cut and as close as possible to where Cochran, Osbourne, and Rourke were when they had their encounters. My camp gear consisted of a small, brightly colored tent—bright red—a sleeping bag, a folding metal table, a battery-powered, Coleman reading lamp, a canvas chair, a five-cell flashlight, a book…and nothing else. The Jeep in which I had driven up the mountain I parked down on the clearing's access road, a good 150 yards away from the edge of the clearing and quite well concealed in thick brush.

I had my tent set up by about six-thirty and then gathered some firewood and got a small campfire going. (Like the red tent, the campfire was to be part of my attraction plan.) I also put a small electric reading lamp on the metal table. This done, I placed my binoculars beside it and settled down to read.

By about eight the light was fading fast and soon afterwards it was dark. I sat at the fire for another hour and a half, reading by the light of my little Coleman lamp, keeping my binoculars close to hand, getting up to replenish the fire from time to time and listening to the night sounds. These included a nearby owl, the croaking of frogs in a pool somewhere outside the clear-cut, and the faraway calls of a coyote pack. There were no other sounds.

At about nine o'clock, I decided to turn it for the night. It was now quite dark, with only a faint starlight illuminating the clearing, on the edges of which a thin, grey ground-mist was slowly gathering. The light was poor but still good enough for me to see all the way to the edges of the clearing, especially if I used my binoculars. But beyond that, the solid walls of forest that surrounded it were jet black and impenetrable. If Mr. B. is around, I thought, he's not around here. Obviously, if he wishes, with his legendary physique, he could be twenty miles away by now. So I decided to call it a night and stood up, stretched, made sure the little fire was safe, picked up my book and my binoculars, turned off the little light and was just about to step into my tent when I heard a stick crack.

The sound came from the black wall of the forest that edged the clear-cut and it was clear and sharp and very definite. I immediately froze for a few seconds and then got my binoculars up and started scanning the wall from where the sound came. As I did so there was a second crack, sharp and vibrant in the damp night air, the second sound, like the first, obviously made and positively identifiable as a large, thick stick being broken.

Within a minute or so there was a third crack and then a fourth and fifth and after that several more and as the sounds continued, and the direction of their source changed, I realized that whatever was making them was circling slowly around me, and around the clear-cut, right to left.

Moving slowly myself, I replenished the fire and then returned to my chair and sat down to watch and listen. I kept my binoculars carefully trained on the dark wall of the clearing, zooming in on the exact place from where each startling crack emanated, hoping against hope to see some movement. But the pitch-black darkness of the interior behind the clearing's wall defeated me; I could see absolutely nothing beyond that. I thought of trying my flashlight. But I knew that it was not strong enough to reach into the blackness beyond the wall, nor did I want whatever was making the sounds to be disturbed and maybe leave; I wanted it to continue what it was doing for as long as possible in the hope that it might eventually step into the clearing itself and let me see it.

The sounds continued for a full hour and though I soon lost count of them, there were, in the end, more than twenty clear, distinct cracks, perhaps thirty.

At about ten, the sounds ceased as suddenly as they had started and after that the night was still and silent. Some clouds came up around eleven, and when they did, the starlight disappeared, leaving the clearing as dark as the forest that surrounded it. I stayed up for another hour after the sounds ceased and then let the fire die and crawled into my sleeping bag and went to sleep.

I left the flaps of my little tent open, tied back, and kept the big flashlight and my binoculars beside me through the night. But when the light of a misty dawn woke me, I had slept undisturbed.

In the morning, before closing up my camp, I did a search of the area around the clear-cut, hoping to find some sign left behind by my stick cracker. But the ground had a thick covering of stick debris—pine needles and twigs—and I found nothing. I searched carefully, keeping a sharp eye on the stick debris. There was a lot of moss and in places it was a six inches thick. There were also dozens of big broken sticks, any one of which could have been used to make the powerful cracking noises I had experienced the night before.

For the sake of my own curiosity, I made one small experiment in the ground debris, crunching through it in my hiking boots to try to replicate the sounds. There was no comparison. The sound of debris being crunched underfoot is distinctive and totally different from the crack of a big stick being broken, in this case, very possibly, by a pair of powerful hands.

If a Sasquatch was close by that night and if it was one of the creatures that made the sounds that I encountered, then of course the question that arises is, why did it do this? Was it trying to frighten me? To warn me of the danger of camping alone in Sasquatch territory? To express annoyance at my camping in its territory? Or was it playing some mischievous kind of game with me, knowing, because of the darkness, that I could not see it?

All I can be sure of is that it was no ordinary animal that made the sounds that I heard, certainly not any of the larger animals of the Pacific Northwest, which are bear, elk, deer, feral boar, and mountain lion. A single stick crack out there in the darkness, made by a hoof, or a paw, maybe. Some debris crunching under a large foot, maybe. But most definitely not the slowly circling pattern of cracking sounds that I heard.

This concludes the account of the two incidents which I feel brought me close to the possibility of contact with a Sasquatch. I am personally satisfied that the bed that we found was made by one of the creatures and that Crew and I made a genuine discovery that day. No man would make a bed of that size and apart from my companion and myself, my guess is that no one had hiked and waded and floundered down that rugged, cold-water, logged-jammed creek in years. A hunter would certainly not go that far to look for game, with the prospect of carrying his meat out on foot and the tiny fish it contained would be of no interest to any fisherman.

Of the two incidents, looking back, I feel that the Fir Mountain experience brought me closer to an encounter with one of the creatures than the bed find. That night on Fir Mountain, if I had had a starlight scope with me, or some other good night-vision device, I might well have been able to see the source of those enigmatic sounds and identify it and maybe even get a photograph of it.

CHAPTER TEN
Origins, History, Numbers and the Reality of the Phenomenon

What I have tried to do in this EPILOGUE is to discuss what I see as essential elements of the Sasquatch phenomenon, such as habitat, sounds, food. I think that I have covered these fairly well here, leaving just four other areas of interest for the Bigfoot enthusiast. These are origins, history, numbers, and the reality—the true authenticity—of the phenomenon. So let's start with origins.

Every living creature on our little blue planet must have an origin, somewhere, at some time. This applies to the smallest of our wildlife, like field mice, and the largest, like elephants. With some, such as elephants, the origins can be surprising, the ancestor of the common elephant being a little herbivore that stood just forty-two inches tall and is scientifically known as the *Moreitherium*. (For more on this, see my Nepal-based books, *THE FIELD GUIDE TO THE WHITE GRASS PLAINS WILDLIFE RESERVE, 2012*, and *THE FIELD GUIDE TO BARDIA NATIONAL PARK*, 2014.) This little fellow was the elephant's true ancestor, with supportive data based, not on myth, but on fossil finds.

We are not sure where the big hominid that Native Americans have named the Sasquatch came from. But we do have the possibility of a genetic origin in a huge primate that lived in south central China at one time. Scientifically called *Gigantopithicus*, if it walked upright—and this has not been fully determined yet—it would have stood over seven feet. That the creature was real—not mythical—is an accepted fact;

hundreds of fossil of bones and teeth have been recovered to date. That it lived in an area that was also the original home of Native Americans, the place from which they eventually migrated across the land barrier of the Bering Strait to come into North America—with the possibility that it came before them, or with them, or not long after them—adds something to the possibility of its being a Sasquatch ancestor and the genetic origin of the creatures of our interest.

As to an historical background to the Bigfoot phenomenon, careful research on the part of associates of mine has unearthed many fascinating accounts of what were very possibly Sasquatch encounters. Some are from journals and letters written by early settlers, missionaries, miners and explorers. Among them, as a few examples, are reports of footprint findings from the British Canadian explorer, David Thompson, 1770-1857; a report of "whistling apes" from the Colville Indian Reservation American missionary Elkanah Walker, 1840; the Yale, BC, incident—a newspaper (The Colonist) story of the capture of a creature called Jacko, 1884; Jack Dover's story of an encounter, as recounted in the booklet *The Hermit of Siskiyou, 1886*; and one of the earliest accounts, a report of the capture of a Sasquatch-like creature by Canadian Indians in Alberta, from *The London Times,* dated 1774; an even earlier report written by the famous German biologist, Georg Wilhelm Steller, which contained his description of a strange, apelike creature seen swimming out towards the ship he was on, while anchored off the BC coast in August, 1741.

NOTE: For more detailed information on any of the above-mentioned explorers and scientists, simply log on to GOOGLE and type in the names. The amount of data available is quite voluminous.

Steller described what he saw as a sea ape. Was it a Sasquatch? If so, what made it swim out to the ship, a distance of possibly a half mile from land? Curiosity? That trait has appeared in other accounts which I and others have investigated, one good example being the Fir Mountain incident, described in the previous chapter.

As to the ability of the Sasquatch to swim, there is little doubt in my mind that they would possess it. It is an accepted fact that the creatures move around a great deal, can be described with some confidence as nomadic—a habit driven by food—and since many of the large rivers of the Pacific Northwest flow east to west, any kind of north-south travel, in line with the lay of mountain ranges, would entail crossing them.

There have been one or two credible accounts of their being seen swimming. One, which I personally unearthed many years ago, was from a learned judge of the Oregon Supreme Court who told me that while fishing in Swift Reservoir, in Washington, he watched two of them swim, together, from one side of the reservoir to the other. The distance across the man-made lake was probably close to 600 yards and they made the swim with apparent ease, emerging from the water within a hundred yards of where he was standing and then, seeing him watching them, walking rapidly away. Steller's creature, whatever it was, obviously had the same capability.

Next, we have the question of numbers of the creatures—how many now survive—and it is one that is easy to approach and discuss because it revolves around a simple if unsatisfactory answer, which is that we don't know. Statements to the effect that there is a precise number, which varies from 100 to 50,000, have of course absolutely no basis in fact and in my opinion are as ludicrous as stating that they are presently to be found in every state in the US.

One man with whom I talked recently—a known name person in the Bigfoot field and actually a sponsor of some research—stated that he was quite positive that at this time there were 12,500 living in the US. I did not want to embarrass him by asking how he knew this. Another informed me that he had information to the effect that their numbers were now down to 250. Asked how he knew this, he told me, with a frown at my look of disbelief, that he got all of his information, on a regular basis, via telepathic messages from "them."

In the end, there is only one number of which we can be sure and that is the one that is a calculation of their biological minimum, i.e., the number needed to procreate and survive. This, I am told by an accredited scientist, would be about 500. So, if there are 500+ out there then they will be able to survive. If there are fewer than this, then they are doomed to extinction.

In conclusion, on the question of numbers and how many are "out there," one long-time, clear-thinking Bigfoot research associate and friend of mine, John Cordell, of Vancouver, Washington, states what is in all of the minds of those who have consciously followed the course of the phenomenon for so many years. This is not how many there are now but, with credible sightings and footprint finds growing fewer and fewer—the last sighting we have now being one in January 2007—if there are any left at all.

As to the reality of the phenomenon, I have no hesitation in stepping forward and saying that I believe in it. I say this of course as a layman—an old field wallah, if you wish—with no scientific qualifications at all and no physical evidence ever seen or found to support my claim, nothing at all...not a bone, no tissue, not an ounce of feces, not even a single hair, not a whisker, as they say. Many claims, I might add, have been made in this area by many people, about having samples of all kinds. But when challenged—put up or shut up—nothing comes forth and the fact is, and I state this emphatically, nothing in the way of physical evidence, nothing, has been found to date.

I personally base my belief in the authenticity of the phenomenon on four areas of evidence which, though circumstantial, have for me solid enough ground. These are the history of the phenomenon, including its Native American background; the eyewitness reports; the footprints—some of them found by myself in remote areas; and the

1967 Patterson-Gimlin film footage, which is discussed at some length in my 1975 book.

Of these, I find the eyewitness reports very real, coming to me as they have from many rational, down-to-earth and thoroughly reliable people, among them state policemen, deputy sheriffs, engineers, surveyors, high court judges (two), school teachers, loggers, naturalists, and experienced hunters. Discarding many of the stories told to me as too insubstantial to contribute to my data base of credible sightings— e.g., fleeting figures seen at great distances—the conclusion I come down to in the end is that all of these people, men and women, saw one of two things: 1) They saw a man (or a woman) dressed in an ape costume, or a head-to-toe fur suit, complete with face mask; or 2) what they encountered was a real, living, large primate of unknown origin and species.

That they did see something, one or the other, I am quite confident. And that it was a man or a woman in a fur suit who has been practicing this risky game—one where, in the Pacific Northwest, it would be very easy to get shot—for several hundred years, certainly seems the less likely of the two possibilities.

In summary of my belief in the reality of the mystery, I personally believe, regardless of how many of the creatures survive today, that they did exist in the Pacific Northwest forests at one time and for a very long time, their inoffensive personalities, their shyness, and their ability to successfully avoid man enabling them to do this with eons of success.

It is 2015 now and many winters have come and gone since I first arrived in the Pacific Northwest to hunt Mr. Bigfoot. Tom Slick, that great man who essentially started it all and who in turn brought all of us into it, has been dead for more than five decades and, with him, many of the fine men and women who worked with me in those early searches. The group who worked with me in northern California, in Salyer and Willow Creek, where we were based in 1960 through 1966, are all gone now, with only their children or grandchildren surviving them.

Tom's dream lives on with me—contact and communication with the creature—but now, I have to admit, it is shadowed by the thought of not actually encountering one before my own run comes to an end. I would certainly like to do so and of course to get a photograph of one, or a bit of video footage. But the clock is ticking.

But at this stage in the game and in the light of Cordell's concern, I can't say for sure that there are still some out there; the dearth of sightings and footprint finds—our main sources of continuing evidence—is beginning to indicate that they might well be gone, which melancholy possibility, I have to admit, bothers me. But I can say, based on the evidence—a strong part of it being the very credible eyewitness reports—and also on the research I describe both here and in my '75 book, that I truly believe that they were here at one time and indeed until quite recently, and that one or two of them may still be here, and that with a little bit of luck, I will get to encounter one before I, too, like so many of my great companions of the Bigfoot hunting years, leave on the last safari.

Appendix

Quite recently, two distinguished professionals have made significant and meaningful contributions of their time and expertise towards a solution of the Bigfoot mystery. These are, Bill Munns, Specialist in Film Making and Special Makeup Effects, of Blue Jay, California and Brian Sykes, Professor of Human Genetics, Wolfson College, Oxford, England.

BILL MUNNS brings a unique perspective to the extraordinary Patterson-Gimlin film controversy. Bill's background is in filmmaking and special makeup effects (such as creature costumes and masks) and he has been active in these endeavors since 1967, the same year the PG footage was shot. So he has an excellent foundation in the film and creature costume processes of the time. A lot has changed since the 1960s and newer researchers sometimes fail to realize that the extraordinary technology they use now was not available in 1967 and that to properly analyze the PG footage one must consider only the processes, materials and techniques that were actually in use in 1967 as a basis for trying to determine if the PG footage is authentic or hoaxed.

With this distinctive perspective, plus seven years of diligent analysis- which includes the accumulation of the most comprehensive film image database on the PG footage and related film material-plus analysis technology and computer graphics which no hoaxer in 1967 could ever have imagined the footage would be subjected to, it is possible to make a conclusive determination about this footage, at least until the argument of whether it is authentic or hoaxed can finally be settled with scientific certainty.

The Patterson-Gimlin Film (abbreviated to PGF) taken in 1967 is virtually unique in the investigation of Bigfoot. Its image clarity is far beyond any other photographic evidence and the subject figure range of motion, posture and activity is ideal for analysis. It is the only photographic evidence which has the potential for conclusive determination. The challenge of a conclusive determination, however, is actually impeded by the tremendous volume of potential photographic evidence, and the parallel vast body of anecdotal and peripheral evidence that has developed by the investigation of the incident and the life of Roger Patterson.

There is a curious dichotomy in the approach to an analysis of this filmed encounter, whereby people who are receptive to the prospect of its being authentic tend to rely on an analysis of the actual film evidence itself, while people who are predisposed to the assumption that the film is a hoax, tend to focus on the character and life of Roger Patterson and argue that because an examination of him leads to many questions about his behavior and integrity-even suggesting the possibility of his being a flawed man-then the film can be dismissed as a hoax. But this approach, as characterized by author Greg Long in his book "The Making of Bigfoot", and then reiterated by author Daniel Loxton in the book "Abominable Science", is to discount the film evidence as useless by examining the filmmaker and not the film. Sadly, this approach upends all classical scholarly and scientific analysis methodology, which cherishes empirical evidence over anecdotal evidence, and relies upon the empirical evidence to advance as far as possible toward a conclusion before factoring in any anecdotal evidence. This miscarriage of what constitutes the best evidence of the PGF has created a controversy that distracts many people from the finest and more reliable scientific analysis of this remarkable film footage.

Early researchers of this film did in fact focus their efforts rightly on the film evidence, but image analysis technology of the late 1960's and throughout the 70's, even into the 80's simply could not reach a proper conclusion because the analysis technology was not yet invented. It was the advent of computer graphics technology and software that finally allowed the film to be properly analyzed and vindicated as a truthful, spontaneous and unexpected encounter with a subject figure that is not a recognized and biologically cataloged species.

A definitive analysis of this unique film footage has finally been accomplished and it relies solely upon the film image data evidence for its determination. The approach used was to firstly look at the physical film itself, how it was photographed, the camera used, the way the film was copied and preserved and, with this evidence, the physical layout of the Bluff Creek, CA. filmsite. The movement of both filmed subject and camera operator were determined in relation to the film frames as the time indicator, and this meticulous analysis indicated that the event truly did occur in a mere few minutes, whereas any hoax would have been staged with breaks between sequences to coordinate the next action. The six filming segments, when the camera starts, when it stops, and the movement of the camera operator while operating the camera, all demonstrate the exact types of actions a person would do in a spontaneous, unexpected and unpredictable encounter, and absolutely none of the actions or movements follow the pattern of staged, planned and controlled filmmaking. This evidence is conclusive because the analysis technology used did not exist in 1967, so no person staging a hoax could anticipate the methods of analysis now used, and thus could not create false positives to fool analysts. The new analysis technology has determined with remarkable precision exactly what occurred in those moments of encounter at Bluff Creek, and the determination is unequivocal that the event was authentic and not staged or hoaxed.

But ultimately it is the analysis of the filmed subject-the creature, so to speak-which provides us the most powerful conclusion and strongest verification that this film footage did indeed capture something biologically real and was not simply a clever performance of a human in a fur costume and mask.

The head shape defies what can reasonably be accomplished with a mask worn by a human. The neck contours do not follow any design for a fur costumes of the era. The contours of the back are remarkable, with true anatomical detail that has no comparable history of use in fur costumes. The breasts have a fluidity of motion which has been scientifically compared to real breast mass and costume prosthetic breast pieces (tested with all the prosthetic materials verifiably in use in 1967 for such costumes) and the breast motions of the PGF subject figure compare perfectly with real biological breast mass motion. In addition, the prosthetic breast pieces tested were a complete failure in replicating the motion in question. The skin-as indicated by the fur growing out of it-has a perfect degree of elasticity, from the torso, under the arm, down to the knee and the right leg stride of the figure in the film has a strong comparative similarity to Neanderthal and Australopithecines, more so that contemporary human anatomy, and the head shape, have a strong affinity to the *Paranthropus boisei* hominid, matching exactly the real skin anatomy of multiple test human subjects. No fur costume material of that era has a comparable elasticity.

The curious contours of the pelvic area, which form patterned clusters of fur, match real human anatomy of people who have substantial adipose tissue deposits on the body, as do the film subject's anatomical proportions.

The analysis of the PGF's subject figure in the film is finally conclusive to an irrefutable degree. It is real. In October, 1967, at Bluff Creek, in the Six Rivers National Forest, in northern California, Roger Patterson filmed, and Bob Gimlin witnessed, a real female "Bigfoot" creature.

BRYAN SYKES. I first became aware of Peter Byrne when I originally set out to research the genetic evidence for Bigfoot and other "anomalous primates" as he so aptly calls them. It took me a year or more before I met Peter at his home in Pacific City where he entertained my wife Ulla and I to a weekend of stories from the days of the great expeditions of the 1950s and 60s that he led. Talking with him, and having read his books, this was an unforgettable experience for me, an ingénue in the field, which added the authenticity that only comes from the lips of someone who has actually been there. And they were cracking stories as well.

From my own point of view, I had, like most people, been intrigued to know whether there was any solid foundation for the ubiquitous tales of wild men and the like, in which I included the *yeti*, Bigfoot, the Russian *alma* and *almasty*. Recent genetic evidence of interbreeding between our own species *Homo sapiens* and Neanderthals, as well as the discovery of several new extinct human species, added a degree of respectability to the notion that Peter's "anomalous primates" could possibly be from remnant populations of Neanderthals or the like.

Viewed from the outside I could also see that the whole field was in a mess, riven by bitter personal divisions, polluted by appalling research posing as legitimate science and a near universal and often expressed sentiment of being "rejected by science". As a professional scientist this was a distortion of the true nature of science, which neither accepts nor rejects anything, but forms a view of the world based on evidence. But in the field of the anomalous primate there was a glaring lack of evidence that would of stand up to the scrutiny required for peer-reviewed publication. There was plenty of assertion posing as science, but that was all.

A few years earlier, my colleagues and I had developed a very robust laboratory protocol for identifying the species origin of a single mammalian hair. The advance was that the new technique overcame the major difficulty of surface contamination with other DNA, usually human, which has always been the major problem in any work on small and damaged samples. Having verified the protocol with a range of museum and other specimens I was sure that this method, based on a sequence of a segment of mitochondrial DNA, would work well enough on the many hair samples that had been collected and attributed to anomalous primates.

To cut a long story short, thanks to the generosity of many curators and enthusiasts in the field, I was able to get a positive identification in thirty cases. With the exception of a Himalayan sample that looks as if it may be from a new genetic relative of a polar bear, all the others were from known mammals. Importantly from my point of view, my colleagues and I were able to publish the results in the prestigious peer-reviewed journal, *The Proceedings of the Royal Society*. As many cryptozoologists would not have access to this academic publication, we arranged to make it available for free open access. *

Although it is understandable that many will be disappointed by the lack of evidence for any new hominid species, that is not my view. All it means is that there were none among the thirty samples tested. Much more important is that there is now a way of providing the evidence of their existence which will be universally accepted. My advice to all concerned is to return to the woods and redouble your efforts to get a hair sample from the real thing.

My own book: "The Nature of the Beast" will be published by Hodder and Stoughton on April 9th, 2015.

OTHER BOOKS BY PETER BYRNE

The Search For Bigfoot, hardback, 1975.

The Search For Bigfoot, paperback, 1976.

Gone Are The Days, 2001.

Gentleman Hunter, 2007.

Hunting in The Mts & Jungles of Nepal, 2012.

Tula Hatti, The Last Great Elephant, hardback, *2005, and Tula Hatti,* the sequel, paperback, 2006.

The Monster Trilogy, 2015.

The Hunt For The Yeti, 2016.

A Fortunate Life, fall 2016

The Hunt For Bigfoot, 2016.

Shikari Sahib, 2000.

Maneater! 2016.

TWO NOVELS, FICTION.

(1) *Rain Falling at Cascade Locks*. A love story.

(2) *The Green Eye*. An adventure story.

TWO NEPAL WILDLIFE PARKS GUIDE BOOKS.

(1) *The White Grass Plains,* The Complete Guide.

(2) *Bardia National Park,* The Complete Guide.

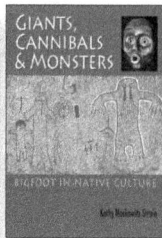